INTERNATIONAL POULTRY LIBRARY

Natural Poultry Keeping

OTHER POULTRY BOOKS

Joseph Batty

• • • • • • • • • • • •

The Silkie Fowl

Old English Game Bantams

Understanding Modern Game

(with James Bleazard)

Understanding Indian Game

(with Ken Hawkey)

Bantams & Small Poultry

Bantams – A Concise Guide

Poultry Ailments

Sussex & Dorking Fowl

Sebright Bantams

Poultry Characteristics—Tails

Artificial Incubation & Rearing

Natural Incubation & Rearing

Domesticated Ducks & Geese

Khaki Campbell Ducks

The Ancona Fowl

Concise Poultry Colour Guide

Poultry Colour Guide

Japanese Long Tailed Fowl

Poultry Shows & Showing

Natural Poultry Keeping

Practical Poultry Keeping

Understanding Old English Game

Old English Game Colour Guide

The Orpington Fowl

(with Will Burdett)

The Barnevelder Fowl

Marsh Daisy Poultry

Sicilian Poultry Breeds

Rosecomb Bantams

Welsummer Poultry

True Bantams

Scottish Breeds of Poultry

Natural Poultry Keeping

Dr Joseph Batty

Chairman
World Bantam & Poultry Society

Beech Publishing House
Station Yard
Elsted Marsh
MIDHURST
West Sussex GU29 0JT

ISBN 1-85736-020-6
First published 1997
New Impressions 1999 & 2002

British Library Cataloguing-in-Publication Data
A catalogue record for this book is available from the British Library.

Beech Publishing House
Station Yard
Elsted Marsh
MIDHURST
West Sussex GU29 0JT

CONTENTS

FOREWORD

This is an attempt to look at the practical problems facing the Poultry keeper who intends to keep a number of birds; whether to go from battery cages, which are destined to be phased out, or some form of free range or related system is the question to be answered.

Many of the arguments against free range are based on extra costs and on the dangers of disease lurking in fields and hedgerows. However, these are based on the assumption that the lowest cost is necessarily the best; in reality the danger to people's health and the welfare of the birds must be regarded as being part of the essential costs. Without examining these the *true* costs cannot be known.

There is no certainty that diseases will increase on free range. What appears to be ignored is the fact that before Battery cages various breeds were kept on free range and the birds, being reared outside, developed anti-bodies which enabled them to resist the ailments and diseases. On the other hand, birds kept in close confinement are subject to the *unnatural conditions imposed by the system.* Fans, insulation, automatic feeding and watering are all features of the factory-type system; unfortunately, the birds are in restricted space, without exercise and if disease really breaks out there is little hope. This system has led to its own downfall with more and more birds being kept in a single cage; greed has resulted in turning the poor hen into a laying machine which is discarded after one or two seasons with brittle bones and very little flesh for the table at the end of it.

The alternative to these battery birds, bred on computer selection, genetically engineered, are the standard breeds, developed in the *fields* of poultry farms. Such breeds as Light Sussex, Rhode Island Reds, Leghorns, Wyandottes and Welsummers, all have records of high achievement, and have won in laying trials, conducted outside. Returns of over 300 eggs per annum have been recorded and, whilst not all can hope to achieve such results, the quality is controlled and specific requirements are covered to suit the consumer. If dark brown eggs are required then breeds like Barnevelders, Marans or Welsummers are the choice; if white eggs are called for (catering) then Leghorns can be kept which should maximize production under natural conditions.

If the consumer is prepared to pay for the privilege of fresh eggs from free range or related systems there seems no reason why the transition cannot take place. In fact of course, if the natural eggs are the main type available competition will ensure that they are at a reasonable price.

J Batty

COLOUR PLATE

The colour plate shows a few examples of the breeds which are available, ranging from the dark-brown egg layers (medium production - Marans, Barnevelders and Welsummers) to the rather poor layer the Indian Game (a table bird), and the Leghorn a layer of white eggs, but a naturally large producer.

Unlike the Hybrid types, these birds have been bred for outside living, complete with the power of resistance against common ailments.

Pairs of Popular Breeds

INTRODUCTION

MANY MISCONCEPTIONS

The description 'Natural poultry-keeping' has been selected because it seems to sum up best the idea of keeping poultry in the most suitable environment for the birds, and the conditions necessary for producing wholesome food. The term 'organic farming' has *not* been selected as an alternative simply because many misconceptions exist on the meaning of this term as used by different people. To some it means adopting a system that never uses artificial fertilizers and only feeds organically produced products; often in practice the border-line between free range and organic poultry farming is quite blurred and in fact, carried to the ultimate conclusion the pure form of organic farming may be extremely difficult to apply.

If poultry becomes ill do we give them raspberry leaves, dandilion roots or some other natural products? Regretfully, we would be faced with the task of culling a great many birds. Even common diseases like coccidiosis, easily cured with modern drugs, would be impossible to control without modern technology.

At the extreme of a system of organic farming the birds would probably revert back to the conditions found in the jungle where the ancestors of modern poultry (Jungle Fowl) still exist, but producing one or two clutches of eggs. It is only by improving the performance of birds and feeding the correct level of protein, carbohydrates and vitamins that the high-producing hen has been created.

If poultry are running outside in a field or paddock, with green grass and other plants in abundance, it can be argued that the natural state is being achieved. Given mixed corn and an ample supply of fresh water the correct balance is being maintained. At this stage however the purist will begin to ask: **How was the food grown?** Moreover, if layers' pellets are given do these contain animal proteins or other 'unnatural' food. Even here there is a difficulty for our free range hen will scratch around and gobble up insects and literally anything small enough to eat that moves! Animal protein is acceptable under natural conditions!

We have to recognize that some form of compromise will be essential; there should be recognition of the pure organic requirement, but we will probably have to admit that some small deviation from the highest ideals will be necessary. ***This is the view adopted in this book trying to give the birds as natural an environment as possible, but also bearing in mind the commercial realities that exist, including the cost of eggs or meat.***

An understanding of the various breeds of poultry, their characteristics and the way food is used and converted to other food is vital to the nature of Natural Poultry–keeping. These aspects will be covered. There is much work to be done to get the standard–bred poultry back to the levels where production is at an acceptable level. The reliance on hybrids, fowl created by geneticists to live and maximize production in cages, should be reduced. Unfortunately, many of the so–called experts on poultry farming seem to think that only the hybrid strains exist and continue to recommend these, even though we are dealing with a different situation.

There must be fresh thinking before it is too late and those that doubt the abilities of the standard breeds to produce a large number of eggs should look to the **Trials** of the 1930s or thereabouts to see how Leghorns, Light Sussex, Wyandottes and others performed under competitive conditions. The re–introduction of Laying Trials on a National scale would bring back interest and enthusiasm to those who take pride in poultry breeds and their sound management.

1

CHARACTERISTICS OF THE FOWL

EARLY ORIGINS

Poultry have existed from a period so long ago that records do not exist. Early accounts suggest that cockfighting may have been one of the principal reasons why birds were domesticated. Why keep birds in captivity when the eggs could be found in the wild and eaten?

Various researchers have suggested that the common ancestor is one or more of the jungle fowl :

1. The Red Jungle Fowl

(a) Gallus bankiva (Gallus gallus)

This appears to be the most likely ancestor of the domesticated fowl, but not all agree on this selection.

(b) Gallus spadiceus

This is the Burmese jungle fowl and modern writers now appear to favour this species (see *Behaviour of Domestic Poultry* , D. G. M. Wood–Gush).

2. Grey Jungle Fowl

This is the *G. Sonnerati* , a species which has been bred frequently in England.

3. Ceylon Jungle Fowl

The *G. layfayettei* is a distinctive species and is largely an orange colour.

4. Green Jungle Fowl

A beautiful green colour and comb without serrations makes the species quite distinctive.

5. Missing Link Jungle Fowl *Gallus giganteus*

This jungle fowl no longer exists and, therefore, this apparent missing link must be based on the assumption that an ancestor existed which is quite different from the other species. Many modern breeds of poultry follow a conventional pattern in terms of anatomy, but there are those which do not : Aseel, Malays, Brahmas, Cochins are examples. It is argued that such large birds could not possibly be created from the jungle fowl known today. There are many physical characteristics which are different from G. bankiva, notably:*

(a) Malay–type birds run and do not fly like the *G. bankiva* type of bird.

(b) Bone structure is different – heavy, strong bones, very muscular for scratching and running. They dig deep holes to find food, unlike the Bankiva which scratches on the surface. Frogs, insects and other natural food is the normal diet of the Malay–type fowl.

(c) Spinal cord arrangement in the skull is **different** in the two types of species:

(i) Malay (ii) Bankiva

For those various reasons it would appear that the Malay–type fowl certainly and possibly other Asiatic breeds came from the *G. giganteus* .

RESULTS OF DOMESTICATION

As a direct result of domestication the fowl has undergone many changes:

1. **Increase in body size**
2. **More productive (increased fecundity).** In the wild a jungle fowl lays between 10 and 20 eggs per season, but the domesticated fowl may lay as many as 300 eggs.
3. **Modified to specific ends** such as brown or white egg production, development of breast meat or other commercial requirements.
4. **Elimination of broodiness** in many races of poultry; e.g. Leghorns and hybrids.
5. **Birds which no longer possess the same degree of vitality or resistance to disease.**

***For a fuller coverage on evolution see *Races of Domestic Poultry* (Sir Edward Brown) and *Keeping Jungle Fowl* (J Batty)**

The Social Obligations

In pushing the birds to the limits in terms of egg production and closing them within an artificial environment, man has created many health problems. These have been combatted by the increased use of drugs in the food as well as by the development of medicines for treating the illnesses which occur.

Poultry farmers have a duty to perform to make sure that birds are not kept in overcrowded conditions or fail in other areas which affect vitality and production. Moreover, it has now been shown that neglect in any aspect of poultry–keeping can result in the creation of diseases which may be transmitted to the consumer in eggs or meat. The incidence of salmonella has caused many consumers to suffer from serious illnesses.

Attention to all aspects of hygiene is absolutely vital and EEC regulations now exist which stipulate minimum standards for cleanliness in production and for dealing with eggs and birds.

CLASSIFICATION OF POULTRY

Poultry belong to the **Galliformes** and may be classified in a variety of ways:

1. **Biological** (Galli)
2. *Standards:*

(a) *Breeds*
 (i) varieties
 (ii) colours

(b) *Classes*
 (i) Hard featured (tight glossy feathering)
 (ii) Soft featured (profusely feathered with much underfluff)
 (iii) Geographic origin such as Asiatic, British, Mediterranean, Dutch, American and so on.

The different breeds and their varieties are recognised by the poultry clubs and breed clubs of the countries concerned. Official *standards* are issued showing the breeds which are recognised and their descriptions. Type of combs may also form a separate classification.

3. **Economic Characteristics**

Poultry may be divided into categories which show economic factors: For example:

1. Layers
2. Utility
3. Table poultry
   (a) Skin colour; e.g. white or yellow
   (b) Weight categories
   (c) Free range or intensive

When birds do not fall into any of these categories they may be referred to as "ornamental" or "show birds". The commercial viability of a breed is the determining factor; Leghorns are also show birds, but they are usually excellent layers as well. Indian Game make good table birds, but their development is slow as well as being difficult to breed, and therefore they would not be a sound commercial proposition.

Hybrids

Hybrids may come into any of these categories, but it will be appreciated that they have been bred for the modern, controlled environment. They have been genetically created to lay more eggs of a specific type or to add more flesh for a given amount of food. They are a combination of different strains or breeds on the basis of economic considerations. Once established the hybrid strain is expected to produce at a consistently high level. This differs from the pure bred poultry where some strains of a breed may not be very good economically.

The improvement of a fancy point in an exhibition fowl may result in loss of efficiency in egg production and obviously must be guarded against. The poultry farmer can use pure breeds, but of the **utility type** rather than the purely show type when some required show feature reduces efficiency.

ESTABLISH BREED OBJECTIVES

It will be advisable to consider the objectives of the breeds to be obtained before stock is purchased. Broadly speaking there are three main categories:

1. Layers – discussed later;

2. Table birds;

3. Dual Purpose (combining 1 and 2)

In particular the colour of flesh and fat will be of great importance.

White Fleshed: Dorking ; Faverolles ;Langshan ; OEGame Sussex; Scots Greys.

Dark Fleshed: Indian Game ; Brahmas ; Orpingtons ; Rocks ; Malays ; Wyandottes.

These are usually regarded as the top twelve breeds for table. They would be reared on free range to produce a very tasty meal.

Houdans

Creve Coeur **La Fleche**

Possible Table Birds
The aim is to produce plump, broad breasted birds.

POSSIBLE ORIGINS OF BREEDS

As indicated the exact origins of the domestic fowl are unknown. This is not surprising because groups of different (sometimes related) breeds came from various parts of the world and were then bred together. It is thought the original source is as follows (but sometimes there is overlap or duplication of two or more countries, and were then bred together):

1. **Asia**

Aseel(s), Brahmas, Cochins, Croad Langshans, Frizzles, Java Fowl, Malays, Modern Langshan, Nankin Bantams, Pekin Bantams, Phoenix, Shamo (Game Fowl), Silkies, Sumatra Game

2. **Belgium**

Belgian Bearded Bantams, Campines, Malines, Braekel

3. **Britain**

Australorps, Dorkings, Hamburghs, Indian Game, Marsh Daisy, Modern Game, Old English Game, Old English Pheasant Fowl, Orpingtons, Redcaps, Rosecomb Bantams, Rumpless Game Bantams, Scots Dumpy, Scots Grey, Sussex

4. **Chile**

Araucana

5. **Denmark**

Landhen

6. **Europe (country not specified)**

Sicilian Buttercups, Transylvanian Naked Neck Fowl

7. **France**

Bresse, Creve–Coeur, Faverolles, Houdan, La Fleche, Marans

8. **Germany**

Kraienkoppe, Lakenvelder, (also spelt Lakenfelder), Vorwerk, Bearded Thuringians (Thuringers).

9. Holland
Barnevelder, Breda, Friesland, North Holland Blue, Welsummer (There are also Dutch Bearded and Crested breeds).

10. Hungary
Magyar, Transylvanian Naked Neck (also shown under Europe because exact origin debatable)

11. Italy (Mediterranean Breeds)
Anconas, Leghorns

12. Japan
Japanese Bantams, Tuzo Bantams,Shamo.

13. Poland
Polish (or Polands)

14. Russia
Orloff, Pavloff

15. Spain (Mediterranean Breeds)
Andalusians, Minorcas, Spanish

16. Switzerland
Appenzeller

17. Turkey
Sultan

18. U.S.A.
Dominique, Jersey Giants, New Hampshire Reds, Plymouth Rocks, Rhode Island Reds, Wyandottes

POSSIBLE ORIGINS OF NAMES

As noted many of the breeds came from far away countries and their true origins may never be found. Usually the breed takes its name from the original place where it was found, but this is not always the case. The following notes may be useful.

1. French breeds often appear to be named after places in France; e.g. La Bresse, Houdan which are districts of France.

2. Spanish and Italian breeds are usually named from the place of origin; e.g. Minorca, Ancona, Leghorn which are place names.

3. Asiatic breeds do *not* appear to exist in their place name:

(a) Cochin (China) are unknown in that country; this breed was originally known as "Shangaes".

(b) Brahmas are no longer on the Banks of the Brah–mapootra river.

Lewis Wright believed the original Brahma in India "may be the missing link" in producing the large strains such as the Chittagons and Malays.

4. Some breeds with place names do not appear to exist in that place; e.g. Hamburghs, or they are given a description which may be misleading; e.g. Polands which do not appear to exist in Poland. Some writers believe the name refers to the crest or "poll".

2

ANATOMY OF THE FOWL

PHYSICAL REQUIREMENTS

An understanding of how a bird functions is vital. It shows the physical requirements in terms of space and the food and other essentials to keep a fowl healthy. The digestive system and reproduction process are also worthy of study because, again, a poultry keeper must know what is feasible for him to maximise results.

A bird's anatomy may be viewed from two aspects:

1. The **internal organs** responsible for providing sustenance and the means of developing eggs in the female bird.

2. The **outward form** upon which the *standard* laid down by the breed club is based.

An understanding of *both* is vital. The provision of appropriate food, water and minerals is vital to success. Unsuitable foods do not provide the essentials and lead to health problems. For the chicks higher protein food develops bone, flesh and feathers and ensures rapid growth.

The outward shape, including the face, beak, crest, neck, body, legs, wings and feathers show the breed of bird. As noted, it is on these features that the *standard* is based. The external shape and appearance also indicate important facts on the type of bird for its selection as a producer.

THE SKELETON AND INTERNAL ORGANS

The internal organs of a bird fit within a structure of bones the skeleton . Knowledge of the principal parts are:

1. Breastbone

The breastbone or sternum is a vital part of the body. It protects the internal organs as well as a foundation for the flesh and muscles

Main Features of the Fowl

which operate the wings.

In describing a bird the *standard* usually denotes the depth of the body and qualifies with such descriptions as "BODY: Back well filled".

2. Wings

The wings provide the means of flying. Their positioning is of vital importance in determining *style* and *posture*: ; when carried high the thighs are revealed and the result is a taller looking bird. This is important for birds with "reach". Many breeds no longer fly; important for selection for free range.

3. Legs and Feet

The legs and feet provide the means of walking, and perching. They should be free form bumps or enlarged scales.

4. Head

The head is mounted upon the neck which is quite flexible. At the front of the skull is the beak made up of the upper and lower mandible. Size and shape of head is of vital importance.

ANATOMY

The functioning of the bird is relatively simple and yet is an incredible process. Food is converted into flesh and/or eggs which, after incubation become chicks which rapidly grow into adult birds, and then at about 6–9 months of age they too become producers, thus repeating the process.

The main parts are as follows:

1. Beak

Food is picked up by a bird and proceeds down the throat into the **crop**, a bag made of skin, and from there it goes on to the **gizzard.**

2. Crop and Gizzard

The crop is the store for food just taken and this passes into a passageway known as a proventriculis (or ventriculus) before passing into the gizzard. The latter is an almost solid organ which masticates food so that it can be digested.

3. Intestines

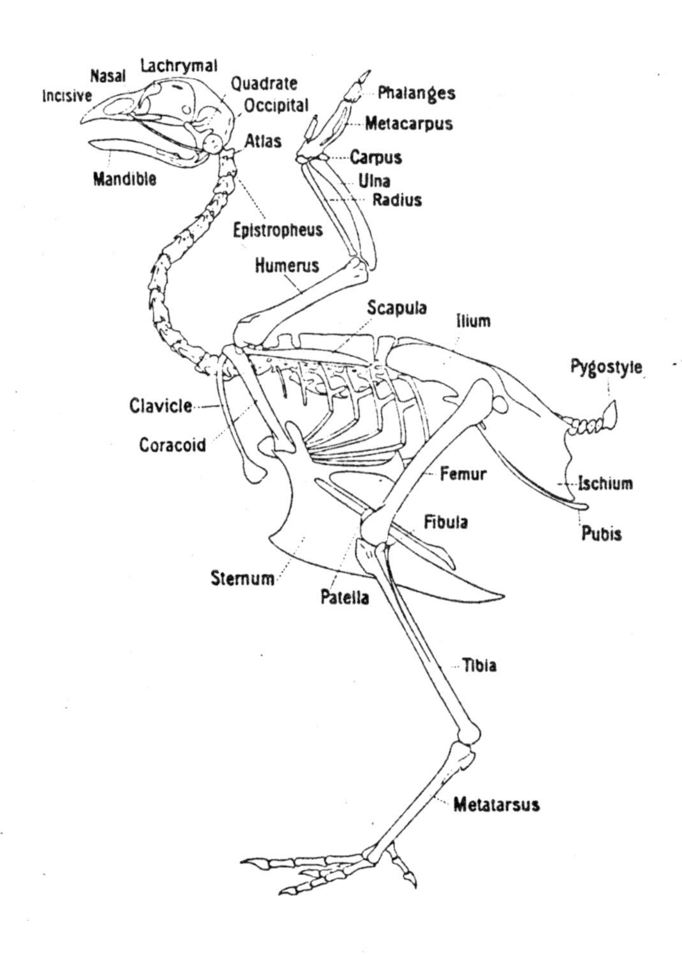

Skeleton of the Fowl

From the gizzard food passes into the intestines and after due processing is passed out through the rectum (or vent).

Within the digestive framework there are:

(a) **Liver** in which is found the gall bladder which stores the bile;

(b) **Kidneys** (two) which filter the liquids and excrete uric acid.

Blood and Air Circulation

The blood is circulated by the contraction and expansion of the heart which is usually likened to a pump which has four chambers – the upper two named "auricles" and the lower two the "ventricles".

The bird breathes through its nostrils or mouth into the bronchial tubes and lungs. The arterial veins from the heart pass through the lungs thus allowing the circulating blood to be oxygenated.

REPRODUCTIVE SYSTEM

For successful reproduction both male and female should be in good health and well fed. Eggs should become fertile within a few days of male and female being placed together, but generally a period of ten days is considered to be a safe waiting time.

The female has two ovaries only one of which usually develops. In addition, there is the oviduct a long twisting tube consisting of two parts through which a yolk passes, adding the various parts ("white", membranes and shell) until the egg falls into the cloaca or egg pouch.

Opinions on how long the process takes vary, but generally around eighteen hours is regarded as the cycle time. Within the ovary there are many embryo eggs (oocytes) – more than 1500 have been counted in a fowl's ovary. These develop so that a few large yolks ripen (usually about five) until one is ready to go into the oviduct for development into the egg.

The male bird "treads" the hen and thereby fertilises the eggs. He discharges semen from testes into two ducts and thence into the oviduct of the female when the mating takes place. For the amateur the fact that a cock need not be run with hens, an important fact in a built–up area, should be stressed. Only if the intention is to breed will the male be essential in a ratio of around 8 hens to 1 cockerel.

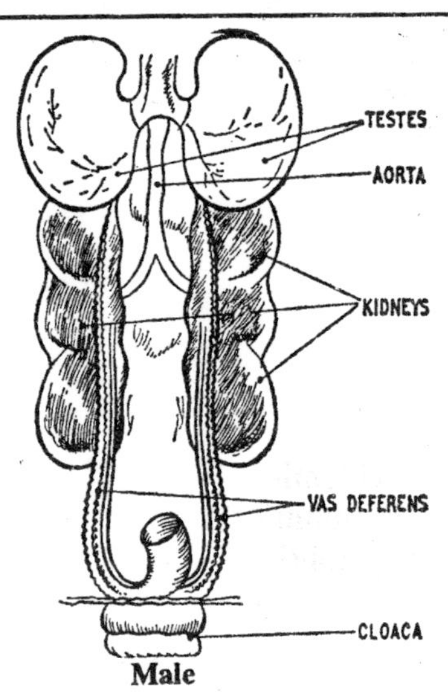

Male

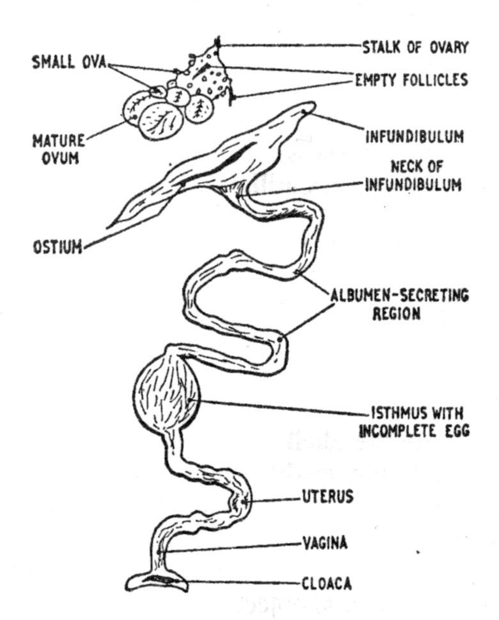

Female

The Reproductive Systems

THE EGG

Egg Colour

The actual colour comes from glands in the oviduct and is transferred by pigments:

(a) oocyanin - Basal Blue
(b) ooclilomin - Yellow
(c) ooxanthin - Red or Purplish
(d) ooporphrin - pattern forming

This colouring stage takes place in the lower part of the oviduct known as the uterus and it is at this point that a coating of calcium carbonate is coated over the shell membrane.

The Shell - General Notes

The outer shell of the egg is made up of three distinct layers:

1. **Cuticle** - a fine coating which gives the egg its lustre or bloom.
2. **Palisade** or spongy Layer - the bulk of the shell (approx. 2/3 of the thickness)
3. **Mammillary** - the INNER part

The shell is very strong and relative to its size can withstand great pressure (Hen egg 60 gramme 4.1 kilos breaking strength, whereas a small fresh egg, 1 gramme in weight, will withstand 0.1 kilo - the Canary egg is around 2 gramme in weight).

Creating the Shell

The top-quality shell comes from the healthy bird, managed in a suitable environment with the appropriate type of food, water and other essentials.

Birds flying out of doors with access to grass, vegetation , earth and other natural objects usually produce eggs of good quality. The calcium carbonate required to produce the shell comes from the food eaten and from sand, stones, leaves and other small items picked up. Limestone is the main ingredient for the shell substance and yet nowhere will it be obviously available.

Proper functioning requires the quantity absorbed to be considered and must be consumed by each hen according to her requirements. Birds kept in cages or aviaries, may be given fine oyster shell and limestone provided in suitable hoppers. If topped up regularly the hens will regulate their own consumption.

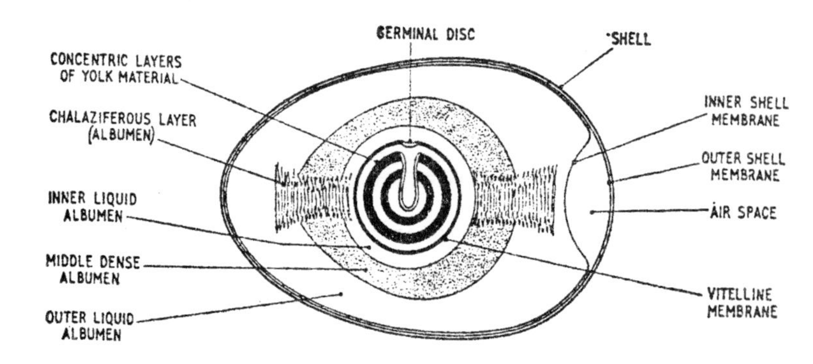

The Egg & its parts

Note: Terms of self explanatory except the CHALAZAE which is a special part of the albumen

In the wild birds eat grit every day even they they may lay only around 30 eggs per season. Obviously, though, egg production is not the only consideration. Adequate nutrition requires an efficient digestive system and this depends upon the functioning of the gizzard. This will function without insoluble grit, but is much more effective when birds are able to eat as much grit as required.

Pheasants, doves, partridges, geese and other birds have been observed taking their daily intake of gravel or other grit. Percentages found in the crops of pheasant was 26 per cent and in Hungarian partridge it was 40 per cent. These were much higher than for grouse (6 per cent) and Mallard (13 per cent. They were nót regarded as conclusive evidence of the normal intake percentages, but rather they confirm the need for regular supplies of grit.

SIZE OF GRIT

When grit is fed to birds it should be **appropriate to the size of bird.** Birds such as grouse or pheasant should be supplied with small granules which they can consume easily in their gizzards should the supply be cut off. Smaller birds should be given grit which is rather like coarse sand. Poultry take pieces about 1mm square in soluble grit,but larger pieces may be picked up , especially the flints used for grinding up the food in the gizzard.

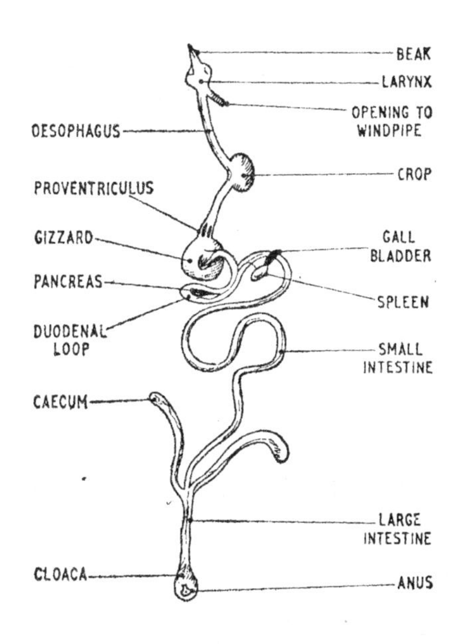

The Digestive Tract

THE MEMBRANE

Within the shell there are two membranes:

1. **Inner Membrane** which surrounds the albumen
2. **Outer Membrane** which adheres to the inside of the outer membrane except at the broad end occupied by the air space.

Air Space

The air space is non-existent in a new laid egg, but gradually appears, taking as little as two minutes (or longer), depending upon the rate at which the egg cools. It supplies a vital air supply to the chick and without it the embryo would die. The older the egg the larger the air space.

Quality of the egg is affected by the freshness and every step possible must be taken to ensure that this condition is maintained as lond as possible. Dirty eggs and subsequent washing, must be avoided because this downgrades any eggs.

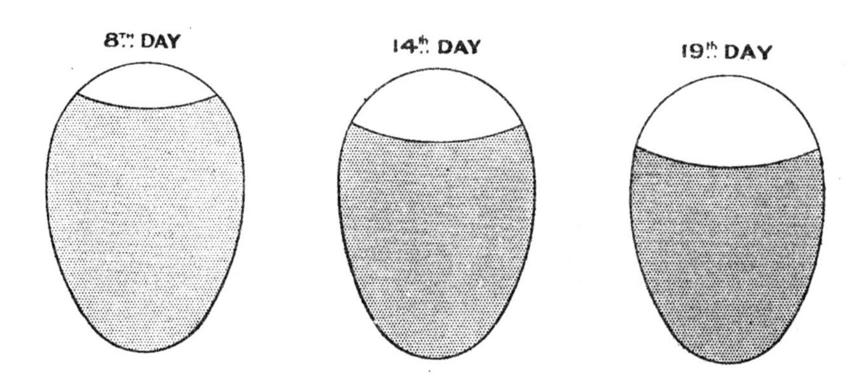

The Air Space

3

THE EGG

NECESSARY ATTRIBUTES

There are many factors exhibited by an egg which should be recognised for successful incubation and for commercial exploitation of the eggs produced for sale.

Factors which tend to be of importance are as follows:

1. **Size**
2. **Shape**
3. **Colour**
4. **Bloom** (the cuticle which should have a glossy finish)

SIZE

The size of a hen's egg is of importance because the larger eggs are usually sold at premium prices. Usually pullets lay smaller eggs and, therefore, are commercially less profitable, but this tendency may be offset by a larger number being laid in the first full year of laying.

There is therefore a differentiation in size between different groups of the same breed of fowl. This may be due to age, but may also be caused by hereditary factors (parent stock) and the physical environment such as housing and food supplied. The ability to convert foodstuff to eggs is an important hereditary factor.

In addition to variations within the same breed there are also obvious differences due to the fact that different species are being considered. Examples are:

		Grammes (approx.)
1.	Domestic fowl	58 – 68

Nest Used for Trapping the hen whilst laying to allow accurate records of laying ability. This also combines a device for allowing the egg to roll beneath the floor thus avoiding the egg being broken or eaten.

Treadle set with door open to admit hen.

Trap Nest

When the hen enters the nest the door closes and she is kept there until released so that her egg can be recorded.

2.	Domestic duck	80
3.	Domestic turkey	85
4.	Guinea fowl	40 – 58
5.	Bantams	38 – 38
6.	Geese (domestic)	200
7.	Pheasant	32
8.	Partridge	18
9.	Peafowl	90

All the weights are approximate and are given as a guide to the relative sizes. When incubating the size is of great importance and generally speaking mixing widely differing sizes together is not recommended.

As noted, in the early stages of production, a pullet lays smaller eggs than in the second and third years. However, after the larger eggs there is then a decline in size which may occur in the third, fourth or fifth years.

The size of the *first* egg tends to indicate the likely size of the "normal" egg to be laid by a hen. A relatively small egg will usually indicate that the later eggs will also be relatively small. Put another way a specific percentage increase is expected which may be more than 10 per cent per annum. Thus:

	Egg A	**Egg B**
Year	*Grammes*	*Grammes*
1	40.0	60.0
2	46.0	69.0
3	50.6	75.9

Body weight also affects the size of the egg. Accordingly, the larger bodied birds tend to lay larger eggs and by the same token birds which have matured *before* coming into lay will tend to lay larger eggs throughout the laying period. *This fact has an important bearing on the rearing methods used.* However, science can do wonderful things and it has been shown to be possible to breed smaller hens with larger eggs within a controlled environment. Whether this is commendable may be debated. A egg which is very large relative to the size of the bird will impose a strain on the system and mortality rates may increase.

SHAPE

The shape of the egg is brought about by the muscles in the oviduct. A smaller egg than normal passing through an oviduct will tend to be spherical.

Heredity factors determine the shape and, therefore, the correct selection of the eggs for incubation should produce layers which will produce similar shaped eggs, provided the cock used is from a strain which also lays similar shaped eggs. The introduction of a new strain may upset the shape – a round egg strain will be changed with new blood which was bred from conical shaped eggs.

Variations in shape within reasonable limits do not affect hatchability. Obviously, though, abnormal shapes will not usually hatch, nor is it desirable they should. Small eggs with little or no yolk come into this category. The double–yolked egg arises from the simultaneous growth of the ova which are then released together. Both size and shape are affected by the presence of double yolks. Theoretically there is no reason why double–yolked eggs should not hatch, but the experience of the author has been that they are not fertile. Obviously they are best excluded from a setting.

EGG COLOUR

The colour of the egg is one of the mysteries of nature. it is only in recent years that ornitholigists and scientists have understood how the colour is formed in the oviduct , mentioned earlier.

There appears to be a connection between country of origin and colour of the eggs: for example:

1. Mediterranean breeds – lay white eggs

2. American and British birds – lay tinted eggs

3. European breeds which lay dark brown eggs; e.g. Marans, Welsummer and Barnevelder.

4. South American breed, the Araucana, lays a green–blue egg

***Recently a hybrid strain was developed by Thornbers which laid a dark brown egg with speckles after the Marans.**

developed strains of small bodied birds which lay brown eggs.

There are breeds which lay brown eggs naturally and they would be more appropriate than hybrids for the rigours of free range. Top of the range for deep brown eggs are:

1. Marans (the eggs may also be speckled with dark brown spots)
2. Welsummers
3. Barnevelders

Others, but not as deep in colour, are indicated in the next chapter.

Barnevelders
Tend to be under-rated as brown egg layers.

White Leghorn : Excellent layers of white eggs.

Anconas: Splendid layers, but wild.

4

BREEDS OF POULTRY++

A wide range of breeds is available , developed for a specific purpose and , therefore , their characteristics should be studied before making a choice. The number of eggs laid, their colour, whether the breed lays in winter and the food consumption should be considered.

TINTED EGGS (Cream to Brown)

Australorp
Barnevelder
Brahma**
Cochin**
Croad Langshan
Dominique**
Dorking
Faverolles
Flemish Cuckoo
French Cuckoo
Frizzle**
Huttegem
Java**
Jersey Giant
La Fleche
Lakenvelder
Langshan
Malines
Marans
Marsh Daisy
Modern Game**
Modern Langshan**
New Hampshire Red
Norfolk Grey
North Holland Blue
Old English Game **
Orloff
Orpington
Plymouth Rock
Rhode Island Red
Shamo **
Silkie
Sussex
Vorwerk
Welsummer
Wyandotte
Yokohama **

**** May only be suitable as show poultry.**

***Some breeds listed are no longer available generally, ie; are rare breeds.**

++See *Breeds of Poultry & their Characteristics* , Batty J for potential as producers

WHITE EGGS

Ancona
Andalusian

Appenzeller
Brabant
Braekal
Campine

Creve Coeur
Friesland
Hamburgh
Houdan
Kraienkoppe

Leghorn
Magyar
Minorca
Old English Pheasant Fowl
Orloff

Poland**
Redcap
Russian Dutch
Schlotterkamm

Scot Grey
Scots Dumpy
Sicilian Buttercup**
Spanish, Black**
Sultan**
Sumatra Game**
Transylvanian Naked Neck**

Some are general purpose birds, others are layers, whereas a few are table birds.

** Likely to be kept for exhibiting only; some others are very marginal.

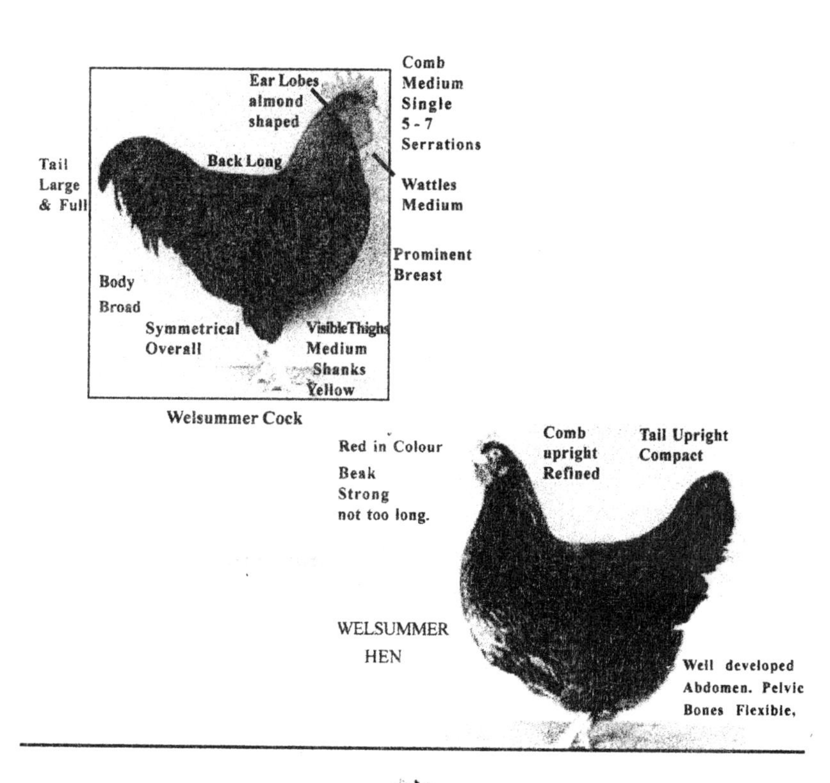

Welsummer Cock

WELSUMMER HEN

MARANS

Typical Brown Eggers

Important commercially because the public believe brown eggs are better.
(They only look more attractive)

Black Orpingtons

Dark Dorkings

Utility Breeds which serve for laying and table.

Indian Game

avy weight. but slow to develop (8 lb or more)

Table Breeds

Some are slow growing and are yellow fleshed so should be crossed to improve growth rate and meat quality , eg Indian Game x Light Sussex .

TABLE BIRDS
Aseel**
Black Sumatra**
Bresse
Creve Coeur**
Dorking
Indian (Cornish) Game**
Jersey Giant (12lb or more)
La Fleche**
Malay**
Orloff
Sussex

Generally speaking the heavier, brown egg layers produce fewer eggs than the Mediterranean breeds and this factor should be borne in mind when selecting the appropriate breed. Also the broodiness no longer exists in certain light breeds, such as Leghorns, and is over exaggerated in the Silkie, a very small fowl, kept mainly for hatching eggs.

BREAKS IN LAYING CYCLE

A hen lays in cycles which are broken by events such as the moult, broodiness, winter conditions and any factor which upsets her health. It follows that changes in feeding , water supplies, temperature and even moving from one house to another can all trigger a halt in production

Any physical problems caused by disease will tend to affect the reproductive organs and, therefore, will bring a pause in production.

The significance of breaks in the laying cycle is that size is affected. Usually there is a decline when the hen feels unwell and, after the pause in laying, the first one or two eggs tend to be smaller than normal.

Maintaining body weight is vital because this affects the size of the eggs laid. There must be a properly balanced diet including the essential vitamins such as Vitamin D.

Usually the first few eggs are best taken away from the hen and not incubated. Obviously, though, with a rare bird this may not always be feasible for only a limited number of eggs may be laid and to continue with the breed all eggs laid must be used.

****Exhibition fowl, but may be used for crossing with others, eg, Sussex, to produce white fleshed table birds.**

5
BASIC SYSTEMS

POSSIBLE APPROACHES

There are many systems available for keeping poultry:

1. Free Range

"Free Range" means the birds have full and free access to the outside, leaving and returning as they think fit. It is the **Natural** method which has many advantages and is regarded by many as being more humane than the other methods.

There are also disadvantages including the cost of land and possible loss of production in inclement weather.

A modified practical definition is considered later in this chapter.

2. Restricted Free Range

Birds may be kept in field units on grass or in pens, but with less land than **full** free range. For those with limited space this is possibly the best compromise, but has its limitations especially on the numbers which may be kept.

3. Semi–Intensive (including straw–yard, barn and aviary systems)

The semi–intensive system means that the birds have access to a grass run, but may be kept indoors when the weather is inclement. In addition, the amount of space available in the outside run is less than free range.

A modification is the **farm–yard** or **straw–yard system**, whereby poultry are allowed to roam around an enclosed yard which is littered with straw.

4. Intensive System

SPECIMEN REGISTER OF RESULTS
MONTH ...January.....................

Date											Total	Average Price of Eggs	Eggs Due to Hatch	REMARKS
1						1					1		Jan. 22	
2		1			1		1				3		Jan. 23	
3						1					1		Jan. 24	
4		1					1				2		Jan. 25	
5					1	1					2		Jan. 26	14 in week
6		1									1		Jan. 27	
7						1					1		Jan. 28	
8											–		Jan. 29	
9											–		Jan. 30	
10											–		Jan. 31	
11											–		Feb. 1	
12											–		Feb. 2	
13											–		Feb. 3	
14											–		Feb. 4	
15											–		Feb. 5	
16											–		Feb. 6	
17											–		Feb. 7	
18											–		Feb. 8	
19											–		Feb. 9	
20											–		Feb. 10	
21											–		Feb. 11	
22											–		Feb. 12	
23											–		Feb. 13	
24											–		Feb. 14	
25											–		Feb. 15	
26											–		Feb. 16	
27											–		Feb. 17	
28											–		Feb. 18	
29											–		Feb. 19	
30											–		Feb. 20	
31											–		Feb. 21	
Total		3			2	4	2	1			12			

Register of Results from Laying Hens

This may be used along with trapnesting (See page 22)

The intensive system means keeping a large number of birds in a limited space. It may be distinguished by reference to the system of management:

(a) regular cleaning and replacement of litter; e.g. shavings, peat moss.

(b) **deep litter** when a very thick layer of shavings is put into the shed and kept for 6 months or longer without being cleaned out. The manure is scratched into the litter and by bacterial reaction is rendered 'neutral'.

(c) **battery system** which entails keeping birds in cages and having food and water provided quite automatically.

SPECIFICATIONS OF CAGES

The cages should comply with specified standards laid down by the regulations of the country concerned; e.g. EEC regulations would apply to the countries in the Community. Birds confined in too small a space, or too many birds per cage can result in lost production as well as unhealthy hens. Bones become brittle, feathers are worn and birds become pale and lacking vitality.

Essentials are as follows:

1. **Reasonable capital cost per bird**
2. **Large enough for birds to turn around**
3. **Bars designed so as to give maximum access to food and water for all occupants and yet not catch or rub feathers.**
4. **A monitoring system to be able to watch over birds on a regular basis**

UNACCEPTABLE SYSTEM

From what has been stated and from the basic definition of Natural Poultry–keeping, given in the Introduction, the reader will appreciate that many systems fall far short of a reasonable ideal and therefore should not be used. The further away from ***Free Range*** we get the more undesirable the method.

Battery cages, even under the regulations which require more space and perches, are still far away from a satisfactory standard.

SELECTION OF APPROPRIATE SYSTEM

The above systems may be adopted in their entirety or in some form of combination e.g. free range for most times, but semi–intensive during the winter. Practical considerations are as follows:

1. Public Opinion

In recent years there has been a wave of feeling which is in opposition to intensive systems or any methods which appear to restrain birds or over–crowds them. Free range eggs may be sold at a higher price which helps to offset the increased costs.

2. Capital Costs

A fully intensive house must be properly insulated and have special ventilation (usually fans). In addition, the cages, plumbing, automatic feeding system, controlled lighting and other requirements all add to the very high capital costs.

Free range systems are cheaper to instal, but the labour costs of feeding and management are higher.

3. Land Availability

Free range requires considerable land and, therefore, when large numbers of birds are to be kept a farm is essential. Imagine the numbers achieved in recent years and the land required! Holdings with 1 million birds have been kept and this is only possible with the battery system.

Opinions differ on the space required for the pure system of free–range poultry management. It seems certain that not more than 100 birds per acre should be kept. The semi–intensive system may allow 300 birds per acre or, with clever management, even more. However, it will be appreciated that any attempt at overcrowding could lead to severe problems (discussed later).

4. Type of Land (See page 75 for Breeds and types of Soil)

Land which becomes waterlogged and muddy is not ideal for poultry farming. On the other hand, dry, sandy scrub–land will not supply luscious green grass. The latter is excellent food for birds and, in effect, is free. Accordingly, its use would be maximised.

Very hot open areas are not conducive to maximum egg production. Trees or hedgerows may provide some shade. The siting of the

The Fold System

Allows access to grass and fresh ground, but is labour intensive.

Barn or Hen Yard.

Stock allowed outside so quite healthy although grass not readily available.

houses will also affect the exposure to the weather.

Some breeds are more suitable than others for heavy soils and these should be selected with care .

5. Access to Services

Poultry require a regular flow of clean, pure water. Accord–ingly, some form of piped system is essential. Hose pipes can be used but there is a limit to the size of range which can be watered. The problems of freezing–up can also be serious.

The foodstuffs, grit, etc. can be transported by tractor and trailer or on a four wheel vehicle.

Lighting is vital and power is desirable so again provision of this service should be considered.

Practical Considerations

Bearing the above factors in mind it is useful to summarise some of the matters to be examined:

1. Domestic Housing

Adequate housing for the poultry farmer and his family will be essential. Adjacent to the house should be outbuildings for the following:

(a) food storage and mixing room;

(b) egg storage;

(c) gate–sales shop;

(d) incubation room (if required)

2. Access to Range

Full access to the range is a vital requirement. Hilly, undulat–ing terrain should be avoided not only because of the difficulties of reaching the poultry houses, but also the inherent problems of security from foxes and thieves.

3. Ease of Partitioning for Field Rotation

As will be shown later, for many reasons it is desirable to rest each area to allow the grass to grow and to avoid infestation with disease carrying bacterias. Careful planning will repay itself many times over.

4. Privacy for Birds

Birds should be kept free from disturbance.

A MODIFIED DEFINITION OF FREE RANGE

Free access to a large area such as a field is the recognised definition of "free range". A population density of 50 to 100 birds per acre gives a guide to what is required. Assuming 100 birds per acre and allowing for a rotation system with changes every 6 months, 20 acres will be required per 1000 birds. For obvious reasons the **pure form** of free range may not be very profitable.

Instead, a *modified* type of free range may be necessary. Essentially the requirement is access to fresh air and a plentiful supply of natural foods without the fouling of the grounds. A great deal can be done to provide the necessary **ingredients** of free range.

The fresh air is no problem so the remainder must be provided in a variety of ways; for example:

1. **Access to grass** with a regular change over of birds to fresh pasture, thus allowing the grass to grow again.

2. **The fold system,** moving the arks across the field on a regular basis.

3. **Pens/aviaries/barns** in which supplements are regularly given by the addition of grass clippings, weeds, leaves and other foundation material.

All these enable birds to enjoy considerable freedom, to scratch and pick up green stuff, grit, grubs and other essentials. Above all, poultry must be kept interested so that they are constantly engaged finding food and enjoying it. Happy birds are productive birds.

The main features of these methods are now considered.

Range Shelter

This method is cheap to produce and is ideal for the summer months.

6

SIZE OF HOLDING

SMALL OR LARGE ?

A fundamental problem with free range is how big is the operation to reach. If relatively small with, say, 100 layers , then problems are not usually serious. Once the intention is to keep 1000 birds **and upwards** the difficulties multiply.

For the **small to medium size** type of operation with one or two fields available the free–range poultry keeping should be able to avoid all the excesses likely to cause problems. These are as follows:

1. **Overcrowding**
2. **Sour ground**
3 **Disease** becoming established in the ground so that each generation of birds has a high rate of mortality or poor production
4. **Disturbances** due to behaviour of birds; fighting, over–lapping of territories and the inevitable establishment of the "pecking order".

OVERCROWDING

Signs of overcrowding will be reflected in the ground becom–ing sour and the grass disappearing – usually in large patches, around and near the poultry houses. Once denuded there is great difficulty in getting the grass to grow. Instead of having natural surroundings to keep them occupied the birds are left to mope around or to stray further and further away with the inherent dangers of being killed by dogs, foxes or traffic.

Sites which are very exposed – a large open field – will tend to suffer from denuding of the grass, especially in summer. Some form of sprinkler system will overcome the problem provided the density of stocking is reasonable. With too many birds per acre nothing can stop the effects of wear and tear on the land.

Accordingly, a shaded spot is better with the sheds and large patches of the grass area being out of the direct rays of the sun. It is for this reason that many orchards have been used for keeping free range poultry. The hens can scratch under the fruit trees and will clear the area of insect pests. In addition, they will keep the grass short and eat any apples or pears which fall from the trees. Obviously though there is a limit to the number of birds which can be kept and the trees may make access to poultry houses quite difficult. For a small number of birds the system is ideal.

Overcrowding leads to stress in poultry. If too many are together or in small houses, but in near proximity, bullying and feather pecking will result with a loss in productivity. Moreover, the poor producers may not be detected with a lowering of the average per flock.

SOUR GROUND

The poultry farmer is faced with a dilemma. He must run birds on the land to the maximum possible extent, thus making the most of his investment and manuring the land to the full, yet he has to observe strict rules of good stock management and hygiene or within two or three years he will suffer quite serious problems.

Steps to be taken to avoid the effects of over intensive stocking are:

1. Move the sheds and the flocks to new ground each year. This is the safest way. However, it does represent a great deal of trouble. A more easily managed system would be to have duplicate housing and after two years to sell off the stock leaving the sheds available for a year for disinfecting, creosoting, etc. and then to restock.

Sheds with wheels or on skids can help if moves are to be made with the poultry houses.

2. Pay special attention to removal of manure and to install–ing methods of keeping faeces separate so the hens do not have to tread in it.

3. Inspect stock regularly and pick them up to check if they are producing. Handling and carrying out simple tests can eliminate the wasters.

4. Watch out for watery eyes, runny droppings, ruffled feathers, fish eyes and other signs of being out of condition.

5. Test eggs and meat of birds at regular intervals to make sure quality is maintained.

6. Watch for bullying and any malpractices such as egg eating.

FREE RANGE PROBLEMS

Psychologically, free range poultry keeping will always appear better than any other system. Fresh air, open spaces, natural food and freedom should lead to healthier birds and, therefore, better eggs or meat. No matter what scientists or poultry farmers state, the idea that ***natural is best*** will always prevail.

In an attempt to show there can be serious disadvantages it is necessary to consider some of the main problems which are as follows:

1. Productivity

Because of the climatic differences and the advent of the moult around 70 per cent production is in the Spring/Summer period and the remaining 30 per cent in the Autumn/Winter period.

With free range there is little chance of improving on this ratio but with batteries, using night lighting, a larger proportion can be obtained in Winter. There is also the creation of a special environment and related requirements such as a full allocation of food, complete monitoring of production and the selection of a temperature which maximises egg production and keeps food costs down. Remember in a cold Winter on free range, a bird will have to eat more to keep warm so production suffers. With the cage system there is no loss because of this fact; the birds are kept warm in an artificial environment.

2. Egg Quality

An argument for free range systems is that eggs are better quality. In fact, scientists have tried to show that there is no difference quality. In fact, scientists have tried to show that there is no difference in **chemical content**, but who will believe this fact when comparing the rich taste of the freshly laid free range egg and the egg from a battery hen !

Analysed into its constituent parts an egg has the following ingredients:

	A %	B %
Water	74	65
Fat	10	10
Protein	13	11
Shell	Excluded	12
Undetermined		1
Ash	1	

Note: A = Free Range **B** = Battery

The differences reflect different sources of information and obviously the methods used would affect the result. Rounding–up of figures has affected the accuracy to the total of 100 per cent.

What was found was that there was no noticeable difference between those eggs from free range and other systems and battery cages.

Provided the food was properly mixed and balanced even the egg colour was similar.

3. Health of Birds

In modern times discoveries have been made which show that hens kept in cages develop brittle bones which may break easily. There is also a serious problem from disease, particularly salmonella, which has affected the acceptability of eggs as food.

Complacency, lack of controls, keeping very large flocks, absence of fresh air and many other factors have been blamed. Fortunately, the danger has been found and the alarm raised. Steps have been taken to avoid outbreaks of disease and to give larger cages and alternative systems to allow each hen more space.

What should also be remembered is that free range can also be subject to disease, which may wipe out complete flocks. Therefore, hygiene and being run on "clean" ground is essential. Food fed to birds is excreted all around and comes with its potential disease, including coccidiosis and worms.

Each pullet will excrete around 200 lb. of fresh droppings each year so imagine the output from 1000 birds on a field. This manure contains about 1.5 per cent nitrogen and essentially will make the grass grow rich and green. Accordingly, if birds are transferred every six months and the ground allowed to "rest", possibly being limed, then the manure can do much good. However, if birds are kept on for long periods without rotation, the ground becomes sour and disease carry–ing, with the resultant deterioration in the health of the hens.

FULL FREE RANGE REQUIREMENTS

As noted earlier, for full range to apply it is essential to have a field on which suitable poultry houses are sited. The hens are let out each day and closed up each evening. They thus have access to a maximum amount of grass which is a valuable source of food and enriches the colour of the yolk to a deep yellow colour.

Security

The poultry houses are usually sited near the dwelling house so the farmer can keep the birds under observation. Opinions differ on how much fencing is necesary. Possibilities are:

1. No special fencing, but place the poultry houses well away from public roads so that birds do not stray and get killed or stolen.

A hedge, barbed wire or electric fence may be quite adequate. Generally houses are separated by about 50 metres so that each flock identifies with a specific shed and with training will go back into it for food and for roosting.

2. Wire netting partitions which allow each flock to be kept separate.

There can be a gate to give access and obviously provided the netting is high enough the birds will not mix. Where cocks are to be kept there should be

adequate protection to avoid fighting. However, much depends on the breed. A pugnacious cock like an old English Game must have boards or corrugated sheets between the pens, but other cocks, such as Wyandottes or Rhode Island Reds will live quite happily together.

Need for Males

If the purpose is to produce eggs or table birds , there is no need to keep any males whatsoever. However, for continuity, so as to be able to produce chicks from the best layers, it may be necessary to have one pen for breeding purposes.

Taken to the ultimate, hens should be "trap–nested" and a record kept of each egg laid. In this way it is possible to know which are the best layers and then, breeding from a cock , **also bred from a top layer,** the breeding pen should produce top layers.

For table birds those with broad breasts and excellent conver–sion properties (turning food into meat) should be the one selected for breeding. Breeds like Indian Game become quite large and plump, but the growth is slow so crossing with another breed such as Rhode Island Red will give speedier growth.

Generally speaking around twelve females to one male **(12 to 1)*** is the correct ratio for breeding purposes. It follows, therefore, that the breeding pen should not be too large. If ten females are used, capable of laying 200 eggs each, this gives a potential hatching of 2000 which, allowing for wastage, is probably a chick potential of around 1700 (a loss of 15 per cent). For most purposes this kind of replacement may be quite adequate.

Against having a policy of hatching and rearing replacements is the danger of disease. Some poultry farmers specialise in producing pullets and obviously this may be an alternative approach. Once re–placements are needed they are bought in to keep production to a maximum.

If this approach is to be adopted then great care should be taken to select a breeder who is known to produce first class birds. Remem–ber too that the hybrid strains developed over the last 50 years were produced for the specific purpose of battery or other intensive man–agement systems. They have not been bred for free range and, therefore, their potential may be suspect. How can we be sure of how such birds will react in conditions which are quite unlike those for which they were developed. It is rather like taking a greenhouse plant outside and hoping it will do just as well. As many gardeners will know this is a very unlikely occurrence.

*** For very selective breeding as low as 3 to 1 may be followed ; eg; exhibition fowl.**

7
HOUSING

FREE RANGE HOUSING

The houses for free range have to be of the type which will give adequate space for normal occupation and possibly an extra area for scratching when the weather is very inclement.

A rule of thumb is that each bird should be allowed 10 cubic ft for roosting purposes ; however it is never as simple as that. Design and positioning of perches, doors, pop holes, nest boxes and food containers all affect the space required.

In the evening, when the birds have gone to roost, look at them in situation and see how much perching space is left. If they are overcrowded with some birds not perching , then it is likely that the design is faulty or the house is too small.

An example of the calculation involved is as follows:

1. Roosting Space by area: L x W x H = No of fowl
Example 6x 4 x 5ft = 120 =12 hens
This is for the conventional poultry shed.

2. Roosting Space by Perch Space

Example as above with perches length ways 18 birds can be kept. Allow three perches with a 2ft space at the front for flying up and down. Perches should be in steps.

This is for an open fronted house

3. Intensive Houses allow 4–5 sq ft per hen of floor space.

This measure is important because it must allow enough scratching space per bird. A house 6x 4ft will house 5 or 6 birds depending on size.

With Barn or Aviary systems the "shelves", mezzanine floors and other means of increasing floor space allow more birds to be kept, but this must be used with caution or overcrowding will result.

A Basic Slatted Floor House.
Better ventilation and automatic cleaning make this very suitable for free range, but tends to be cold in winter.

RUNS

Runs are an essential part of poultry keeping.They supply fresh air, exercise and general interest which cuts out boredom.

Green food, leaves and other additions, provided each day,ensure they are kept active and eager to forage around. A guide on space is :

1. Grass land such as fields , paddocks and orchards.

Allow 100 sq. ft for each bird. A plot of 50 x 50 ft would take 25 hens.Opinions differ on the optimum number , but around 50 maximum per flock is advisable.

2. Gardens

Allow about 20 sq. ft per bird as well as the allowance for the shed.

3. Intensive Free Range

As many as 400 birds to the acre may be possible for relatively short periods, but movement to a different site will be essential every 6 months or so depending upon conditions. Half this number is possible on ploughed land.

If on a permanent basis not more than 100 per acre is advisable.

These measurements are a guide only. The type of soil , time of year , weather conditions and other factors affect the stock intensity.

In winter, when the weather is inclement, to keep birds in lay it may be advisable to keep birds indoors, thus conserving heat . A scratching shed with peat moss, leaves, shavings or other litter is essential. The system the becomes a 'Barn' system and not strictly free–range, but this is better for the welfare of the birds.

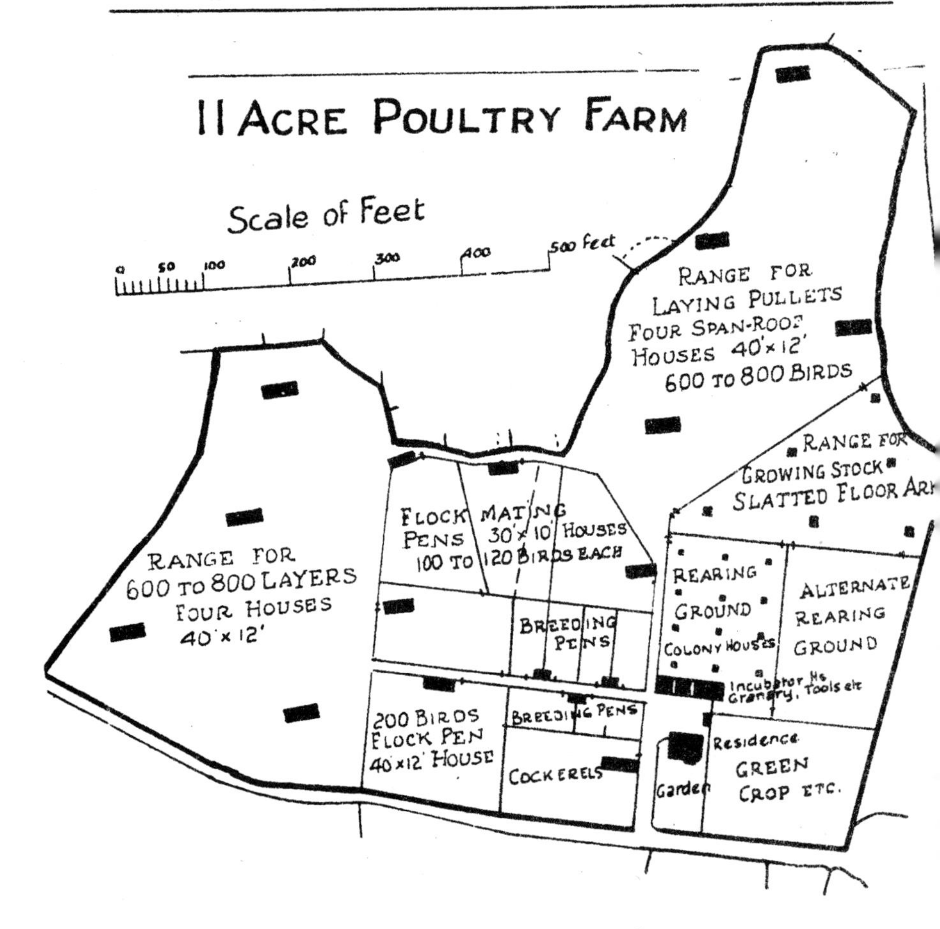

Layout of small poultry farm

The important factor is to keep changing the birds around so they are on fresh ground every other year. In recent times it has come to be acknowledged that 'Integrated Farming' is advisable, ie, running a different type of livestock, such as sheep, with the birds and they eat what is left by the fowl. There is nothing new in this concept; ostrich farmers in South Africa have done this for generations (see *Ostrich Farming* , J Batty, BPH). There is also a safety measure with integrated farming allowing a different product to be available, without absolute reliance on poultry.

Housing breaks down into the form of building used, for example:

1. Poultry sheds of varying sizes to cater for the size of flocks being kept.

There is conflict between the size required for centralised control with large houses holding, say, 3,000 birds and the small unit for 100 layers. The latter gives a more realistic grazing area, but is rather costly when it comes to providing food and services. For large scale operations the large house will be essential because automation will be vital. A stockholding of 10,000, allowing for rest periods for the grass, should have a 100 acre farm, but in practice this size of flock is being housed on 24 acres which in the long run will lead to overcrowding*.

The *Sussex Ark* has been known for generations.

2. Fold units which give free access to grass and yet do not require fences.

3. Special Houses, usually for the larger poultry farmer, e.g.

(a) slatted floor houses;

(b) polyproplene/plastic/fibreglass units. These new materials offer tremendous scope, but have not been tried fully. They are easily maintained, but do not 'breathe' like wood.

SMALL SHEDS

These come in various shapes and sizes. Generally, units should be reasonably small to avoid large numbers in one shed so that bullying and other problems are avoided. The stocking rate recommended is generally three birds per square metre.

If they are to be moved regularly then they should be provided with skids or wheels, thus allowing free movement without placing too much strain on the shed.

Every part of the shed should be used and, therefore, provision should be made to have nest boxes about floor level and food troughs above these.

Similarly, water should be provided in water fountains in each shed or outside. The advantage of the latter is that there is less danger

***Based on reports in** *Poultry World*

of water running low. The founts should be checked regularly, at least once a day.

THE SUSSEX ARK

One of the best known poultry sheds for the small scale commercial operation is the Sussex Ark, said to have been named because its shape resembles the biblical ark. Its main features are :

1. Low being no more than 5ft at the eaves , thus being economic to build.
2. Two doors --one in the top and the other at the front.
3. Slatted floor (See diagram) so that droppings fall through into a cavity.
4. Slatted floor can be removed for ease of cleaning.
5. Ventilation is at the eaves but also through the floor area.
6. Construction is by means of overlapping boards.

The unique natural ventilation enables birds to thrive and remain very healthy and can be employed in a colony system without difficulties once the birds have become accustomed to their own ark. For layers it will be necessary to put a three-compartment nest box for up to 15 layers.

Sussex Ark

Size; 2 x 1 x 1 m x 5ft high at ridge. A colony of 8 arks will house 100 layers with a shed to store food.

PENS AND ENCLOSURES

In suitable conditions the stock is let out into a field each day.Ideally a flock should have a plentiful supply of fresh grass.

Because of predators, birds must be locked away each evening. There is also the need to train them to become accustomed to a particular house so that they will return to it to roost each evening. **Suitable perches and dropping boards must be provided (see examples in text).**

An alternative to the **open-field approach** is to have separate pens to accommodate, say, 50 layers. Wire netting partitions or other barriers are used with a gate for access. If breeding is to be allowed, with cockerels running in the pens, it will be necessary to board up the lower part of the fence so that squabbles between the males will be avoided.

SOIL SUITABILITY AND 'BURNT GRASS' EFFECT

For free range to be successful, attention must be paid to the type and condition of the soil. A well drained, light loam is ideal so that mud is kept to a minimum; if necessary small pebbles or gravel can be placed near popholes and the door so that mud is not carried indoors and possibly onto the eggs.

The **Burnt Grass effect** can be avoided by putting birds on new land at regular intervals with sufficient time to allow new grass to grow. There is a conflict; **low** stocking will save wear on grass, but will not provide adequate use of the land, building and equipment.

Always there is a tendency for birds to stay within easy reach of the house where food and water are available. Thus with the large house the existing grass will quickly disappear and the question is what next? Movement of the stock is the ideal, but cannot always be done very easily. The provision of grass clippings from lawns, paddocks, etc., may provide a solution and the dumping of leaves and similar foraging material is desirable, provided always that the litter is edible and no danger to health.

A variety of houses for the small producer
All are suitable for free range.

Fencing of some kind, preferably easily moved should be used to keep the birds within restricted areas. If there are serious problems with land becoming sour the answer may be to plough the field and re–sow with a quick growing grass. At one time regular liming was advocated, but a more specialised chemical to exterminate disease might be appropriate.

Birds which are quite hardened to outside living are essential because they will have the necessary anti–bodies to resist disease such as *coccidiosis*. . Generally speaking, if the soil is on the heavy side a sturdy fowl such as Light Sussex or Rhode Island Red would be selected. The lighter breeds, which include the Ancona and Leghorn, thrive better on the light soil, but these can be rather flighty and some form of "taming" is desirable before being allowed outside. However, even some light breeds may be all right on heavy soil (see p 34)

Certain hybrids may also be suitable, but proof of outside rearing should be sought from the supplier.

The Rhode Island Red

Utility breed suitable for most types of application. This is a hardy, general purpose fowl with a good laying record.

The use of lights to give the required lighting pattern (e.g. 14 hours per day) for maximizing egg production. However, some would argue that this stimulates the birds and therefore is 'unnatural'. Others would assert that limited light (time and density) simply gives more feeding time which is essential for achieving a reasonable level of production. There is merit in this argument because without light it may be very difficult to give the degree of management essential for efficiency.

The type of housing will allow affect lighting and other services, including the provision of water and food. With a large unit these services can be provided easily, but with small units (sheds scattered across a field) feeding and watering in winter becomes quite onerous and the use of a tractor and food and water transport (a special trailer) will be essential. Yet the verey large unit is quite 'unnatural' and is not in the best interest of the birds.

Stocking should be realistic, but some producers attempt to house at the rate of 10 birds per square metre which may be regarded as too high in most situations; a figure of 3 or 4 is more realistic. Regular change of ground or some form of corrective action such as rotovating is essential and excessive stocking only aggravates the position. Obviously coping with high rates of stocking calls for efficient stockmanship and management.

ROTOMAID
Egg Cleaning Machine

If eggs are to be clean and fresh there should be regular collections and if on a large scale a conveyor system will be necessary, thus minimising the handling costs. Washing facilities for eggs will be essential (although frowned upon by many because of possible misuse contaminating the eggs) and this may be a sophisticated machine or a bucket-cleaner such as the well tried *Rotamaid* system. When eggs are to be hatched there should be *sanitization* by passing the eggs through a special cleaning machine thus avoiding many problems and diseases which arise from dirty eggs.

Ventilation is vital irrespective of system.
This is natural ventilation from the window, suitable for the small producer.

FOLD UNITS

Fold Units are small units consisting of a house at one end and a wire netting covered run at the other (or in the middle). They are easily moved around so that a fresh patch of ground can be used when grass becomes eaten or soiled.

Construction

Folds should be movable on wheels or skids. Sometimes handles are placed at each end. Obviously folds should be robust, but quite light.

In a typical unit there is sufficient room for 20 hens. More than this number is not to be recommended.

Provision must be made for water and food containers and these must be under cover or the feed will be affected by inclement weather. This covered area can be at the opposite end of the sleeping compartment.

PRACTICABILITY

Fold units are excellent for rearing chicks and growing stock. They may also be used for layers and broilers.

The main **disadvantages** are:

(a) Fairly high capital cost to build the units;

(b) High labour costs to supply food and water and collect eggs.

A full-time poultryman would be required to look after around 1,500 birds, but more or less depending upon circumstances. Each one has to be supplied with food and water and usually there is difficulty in introducing an automatic system.

The usual recommended stock intensity **at maximum** is not more than 200 per acre and this would require many fold units. However, it appears obvious that anyone requiring to keep large numbers of birds should think of one of the other systems.

VERANDAHS / SLATTED FLOORS

A variation of the fold unit is the **raised** verandah where birds are kept on wire or slatted floors. The advantage of the method is a plentiful supply of fresh air with the avoidance of any birds being given direct access to the ground, which avoids disease, especially with turkeys.The conventional slatted or twilweld floor allows birds to go indoors to roost , lay and shelter , but with automatic clearance of the droppings , which fall underneath the unit and can be removed quite easily.

Field House on Wheels

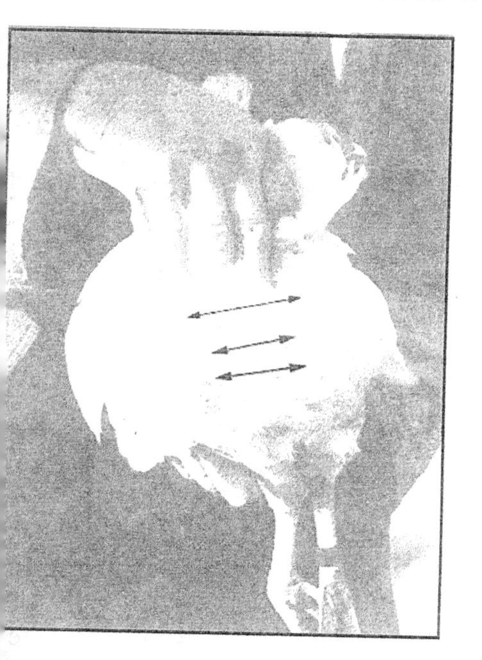

Laying Test

Handle birds regularly. The space between the pelvic bones indicate laying (3 fingers) for bird in production.

SPECIAL HOUSES

Many new ideas have been put forward for poultry houses, the idea being to try to automate as far as possible, yet with the advantage of access to grass and fresh air.

One of the first priorities is attention to labour saving methods so that wages are minimised. Possible areas of attention are:

1. **Ease of Cleaning**

Slatted floors (or twilweld) allow droppings to fall through and, therefore, save labour. If contained in a compartment underneath the slats the ground is not fouled with the droppings, although after a time, if movement is necessary, there is a disadvantage in that the unit becomes quite heavy (regular clearance is therefore advised).

2. **Water**

Filling individual water fountains each day can be time consuming chore and, therefore, a header tank or something similar is desirable. Failing that it is advisable to use a hose pipe.

3. **Foodstuffs**

As for water – some form of automatic feeding is desirable. As a minimum a large hopper would be essential, with a self–filling principle. Grit must also be given.

EXAMPLES OF SHEDS AND LAYOUTS

A number of illustrations are given to show the types of houses and layouts that are found in practice. Some of these are from the days when free–range was a very popular system, before being replaced by laying batteries. Accordingly, they will serve as a guide to those who seek to establish free range.

The *Polybuild* concept attempts to give full automation within a single building. This is flexible and is a fully insulated building. Obviously, like all free range houses, it will be necessary to allow sufficient ground area for the hens to have a regular supply of grass."Burnt up" patches and fouled ground have to be avoided but this should not be difficult if the house is placed in a convenient spot

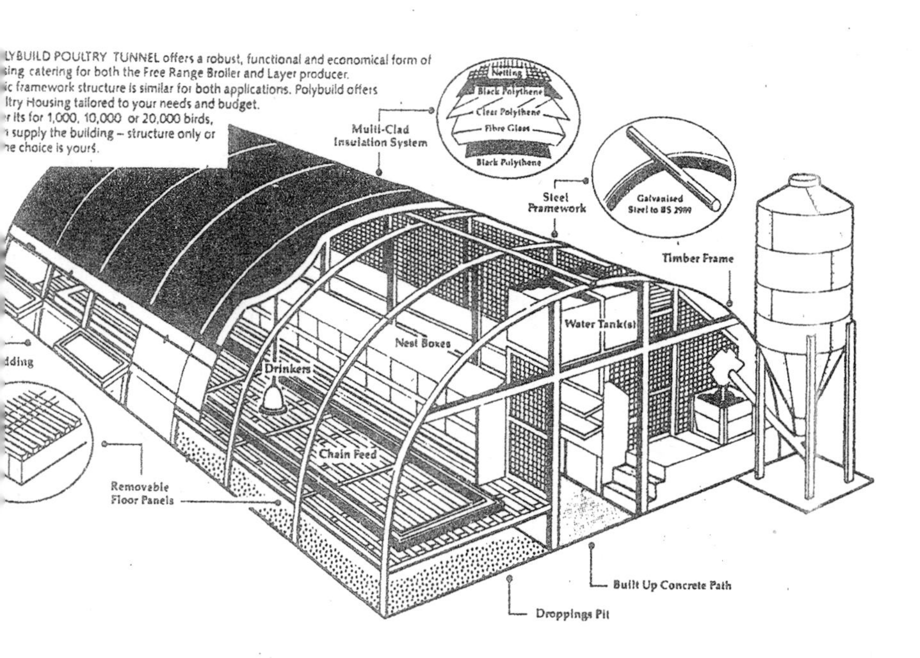

The *Polybuild* free range shed.
Gives flexibility for varying size flocks on free range .

ree Range House for 250 layers The *Skidder* is moved on skids.
Courtesy:John Price & Son

The *Skidder* made for moving around to fresh pastures.

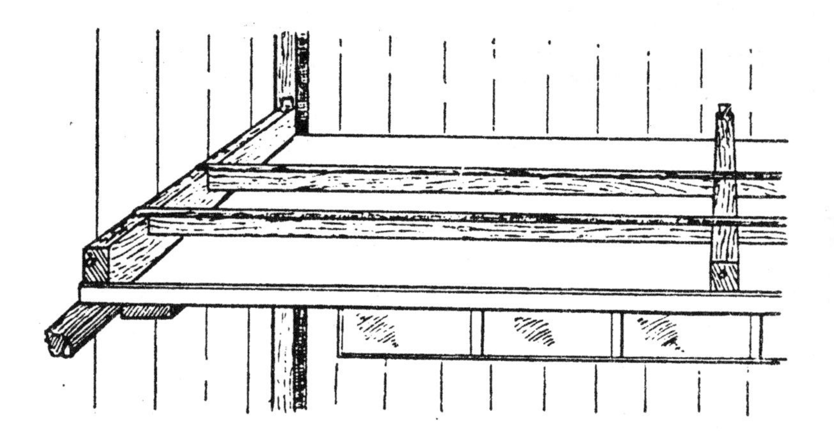

A Droppings board and double Perch

This allows more scratching room and the litter is not too fouled with droppings. The perches are removable for ease of cleaning.

for dividing up the field so that different parts can be used, thus resting sections for a time.

There are claims that the movable house is better and a house which allows movement and yet is large enough for 250–300 birds is now being used. An example is the *Skidder* , a unit 10ft x 6ft which allows ready movement on to fresh ground.

ADVANTAGES OF FREE RANGE

There are many advantages claimed for free range:

1. Birds enjoy natural or organic living with access to fresh air and grass and other natural foods;
2. Eggs are of a high quality with deep yellow yolks;
3. Marketing is easier – farm shop sales or contracts with shops;
4. Lower capital investment.

DISADVANTAGES

Problems which can occur in free range are:

1. Loss of birds through disease.

One of the main reasons is probably the fact that birds being used have been bred for intensive conditions, whereas standard breeds such as Leghorns and Light Sussex are able to withstand exposure to the earth and its potential hazards. Normally mortality under free range is no worse than under intensive methods and can be better;

2. Need for skilled stockmen (or women) who are often in short supply;

3. Dirty eggs and, therefore, the cost of washing these.

This can be overcome by having gravel around the houses or by

Free Range in a conventional house

Around the house is a large area of gravel which can be washed down and disinfected, thus avoiding mud being carried in and possibly contaminating the eggs. Sliding windows provide ventilation and the nest boxes are outside for each of collection. The grass is in good order (no burnt patches) thus denoting there is no overstocking at present.

A large House with Separate Sections

The house is divided into a number of separate self-contained units with pens attached. This is useful for breeding, layers or table birds. It may also be used as a rotation system to allow a section and run to be rested and new grass grown before introducing new stock. If a large house is used there can be a centre corridor with sections on both sides and runs on both sides of the house.

moving houses quite regularly so that mud is not present. Also shavings can be put in the houses or a special canopy covered section can be outside so that eggs as laid roll into a gully ready for collection;

4. **High cost of producing eggs** under free range and the low production in winter;

5. **Birds may eat more food to keep warm.**
In fact, with access to grass birds may eat less. Statistics suggest that around 120 grams per bird per day (4.25 oz) is the norm, but up to 7 oz may be eaten by a prolific layer producing 200 eggs or more per annum.

CONCLUSIONS

Undoubtedly, there are distinct advantages and disadvantages with Free Range. The fact remains that with the ever increasing pressure for **natural farming** the system appears set to become well established.

A word of caution! The words "free range" are capable of many interpretations. EEC regulations apply and no attempt should be made to describe eggs as free range unless they are from birds which do come from a proper free range system. At the time of going to press the exact terms to use to distinguish the different types of production are under consideration; if birds are really on free range with full access to a field, orchard or other open space then the term *free range* may be used. However, if birds have restricted access to the outside or are kept in some other type of system the appropriate description must be given so as not to mislead the consumer ; eg, Barn eggs.

A factor which should also be stressed is that all eggs should be **absolutely fresh** -- too many producers with farm shops pay too little attention to quick turnover of stocks and continue to sell eggs which are beginning to deteriorate. It is essential that eggs are kept **cool** (not frozen) and a special room which maintains a temperature of around 15^{0} C is desirable. Only wash when essential because this down grades the eggs and machines can spot any eggs which have been washed.The eggs should be date stamped to show when laid or freshness date.

8

THE PSEUDO FREE RANGE SYSTEMS

SEEKING AN ALTERNATIVE

In an attempt to find an alternative to the pure free range systems which are costly to operate, various alternatives have been put forward which pretend to be the near equivalent of the natural method. These can be criticized because birds do not have access to pasture or similar open 'grazing' and neither do many of the systems have adequate space or ventilation.

The more enclosed and **intensive** is the system the further does it fall short of being free range. Stocking rates with the Aviary Method may approach the battery/ cage system and this overcrowding in itself disqualifies it as coming under ***Natural Poultry Keeping***. There is no denying the improvement over cages,but there should be no claim for equality with free range.

Even some 'free range systems' , using very large poultry houses with heavy stocking rates , lighting systems (because the shed has no natural light),and various other features from intensive poultry keep–ing , are not really proper free range.

SYSTEMS BETTER THAN CAGES

The major systems available as alternatives are as follows:

1. Aviaries or Percheries

2. Hen Yards or Barn Systems

Aviaries or Percheries

There is no standardized definition of this method. The main feature is the fact that stock are free to mix and use the communal facilities for feeding ,water, scratching and laying. The degree of access to fresh air varies and obviously , despite the freedom , bullying and other vices may occur because the hens are unable to avoid the

The Barn or Hen Yard System.

In the alternative methods this is probably the best one; it gives outdoor access and adequate fresh air.

aggressive types.

Compared with cages all give more freedom for the birds and attention must be provided for the following:

(a) Scratching areas;
(b) Variety so that birds do not become bored;
(c) Eating and drinking facilities;
(d) Perching and resting;
(e) Nesting;
(f) Belt for removing droppings.

A major feature is the provision of tiers consisting of shelves, perch areas, conveyor belts and similar features all of which give more "space" within a given area. The building of mezzanine floors allows more floor space to be created.

Laying hens no doubt appreciate the exercise obtained from jumping up and down. However whether this is really giving birds a more natural environment is open to question. It is an improvement on the cage, but is virtually a large cage and just as clinical.

Stocking of all systems requires careful attention or birds will congregate in one or more parts, leaving the remainder under–utilised.

Perches should give easy access to nest boxes and stacked off the floor areas.

Food consumption can be controlled, but tends to be higher than in cages. Around 130g per day is usual. Free range tends to have the highest rate because conditions are unpredictable.

Vital Statistics

It will be necessary to set realistic standards of performance which will be used as a basis for planning and control.

Usually a flock of 100 birds or more would be taken because too small a sample would invalidate results. For comparison to be made each flock being tested should be kept under identical conditions, thus minimising the effects of the environment.

Information collected may be varied on the basis of the purpose but usually factors considered would be as follows:

1. Egg production

2. Feed per dozen (or 10) eggs
3. Feed per day per bird
4. Size of egg expressed as a percentage;
   e.g. 60g eggs 40%; 50g eggs 15%
5. Mortality (a complete loss!)

Straw Yards/Barn Systems

There is nothing really new in the world of poultry management, so far as they have been developed in the last 25 years. What happens is that the basic systems remain the same with improvments to methods or materials used. Such is the case with straw yards and barn systems which were developed and then discarded in favour of battery cages.

Straw yards operate on similar lines to deep litter in so far as the insides of the buildings are concerned. Here, however, a droppings pit is commonly employed over which the perches are placed. A further 1 sq. ft. per bird of dry scratching area is very useful. Feed and water should be indoors in bad weather, and must be before the birds when under night lights.

Considerable progress was made in designing suitable systems, but then came to a standstill; they are now back in favour. Accordingly, because free range is the vogue the similarity in the two systems is being stressed.

There is much merit in the Barn system, but eggs must be sold as coming from that source and not 'free range'.

9

WELFARE OF THE HEN

MAKING LIFE MORE TOLERABLE

In recent years more attention has been paid to the cruelty aspect of keeping birds in cages. In summary form:

1. **Practical Standpoint**
Hens are not humans and do not suffer in cages provided there is adequate room for feeding and exercise. Moreover, the battery system provides eggs at prices the consumer is willing to pay (they have to be 'educated' into the need for higher prices).

2. **Welfare Supporters**
No animal or bird should be restricted too much; all should be free to enjoy green fields, fresh air and unlimited exercise.

Unfortunately, many of the welfare supporters would object very strongly if asked to pay the premium prices expected for free range eggs. Accordingly, in the long run many experts are suggesting some form of compromise will be essential and regulations will ensure that where battery cages are used they comply with specified minimum standards. This is not a sound argument when viewed from a welfare angle, but may be so on economic grounds.

No matter what adjustments are made to cages, with perches, more room or other improvements, they must still be regarded as a commercial device for exploiting the hen and whether they can be justifiable at all seems very doubtful. However, some of the semi-intensive systems are more humane and acceptable. At least in these alternatives, even if not fully free-range, birds do have freedom. In any event, in inclement weather, just like humans, birds must be provided with shelter.

FREE RANGE HEALTH REQUIREMENTS

If the stock on free range are fed well and moved on to fresh ground on a regular basis there should be no serious health problems. The main weaknesses will arise from a breakdown of the system , either partially or fully. Some of the areas to watch are as follows:

1. **Sound stockmanship;** observation of the birds , taking action in good time, where necessary , and ensuring that food and water is always available are the main requirements. Any unproductive birds must be noted and if there is no improvement they should be culled. Signs, such as a purple comb, may be a sign that the water supply has dried up. Investigate and discover the cause or production will suffer.

2. **Avoid overcrowding** and watch out for feather pecking and other vices. Make sure birds do go outside and enjoy the fresh air and natural food in the paddocks.

3. **Remove any accumulation of faeces on a regular basis and maintain a high standard of cleanliness.** Cleanliness is essential and yet is often neglected. See number 8 below.

4. **Nest boxes should be kept under regular observation** and soiled litter replaced. Dirty, cracked or other faulty eggs are a direct drain on earnings and must not be tolerated. Selection of sutable nest boxes is an important management decision.

5. **Sheds should be appropriate and should be cleaned , painted or creosoted and kept in good repair.** Adequate ventilation is essential, but at the same time for winter conditions, sound insulation is advisable for a hen will not lay if almost frozen to the perch each night.

FEATURES.	GOOD LAYER.	POOR LAYER.
Head . . .	Well balanced, flat on top and broad.	Long and narrow, or rising sharply in front and falling behind.
Eyes . . .	Full, bright, prominent, set well up in face.	Small, sunken, dull, set low and far back in skull.
Face . . .	Lean, good colour, fine texture.	Fat, coarse-skinned, or pale and thin.
Comb and Wattles	Fully grown, red and warm, fine silky texture.	Shrunken, dry and dull.
Body . . .	Broad back, deep breast, straight, medium length breast bone.	Narrow weedy type, or coarse type with round thick bones.
Shanks . .	Thin and flat with smooth scales.	Round and fat with coarse scales.
Feathers . .	Tight and well worn, late and quick moulting.	New clean feathers, loosely carried, early slow moulting.
Skin . . .	Soft, silky and pliable.	Coarse, dry and thick with hard layer of fat beneath.
Pelvic bones .	Prominent and pliable.	Thick, stiff or covered with fat.
Distance between pelvic bones and breast bone .	If laying, 4–5 fingers breadth between pelvic and breast bones, 2–4 fingers between 2 pelvic bones themselves.	1–2 fingers breadth between pelvic and breast bones, pelvic bones close together.
Vent . . .	If laying, large, moist and pale.	Small, contracted, round.
Yellow colour (in yellow - fleshed breeds)	Entirely absent in vent, eyelid and beak and shanks	Present in vent, eye, beak and shanks.

Cull poor Producers

The abdomen test is a sure sign of a laying hen. Two fingers are place in the gap between the bones; if no gap the bird is not laying. Condition as indicated by comb and eyes are also signs.

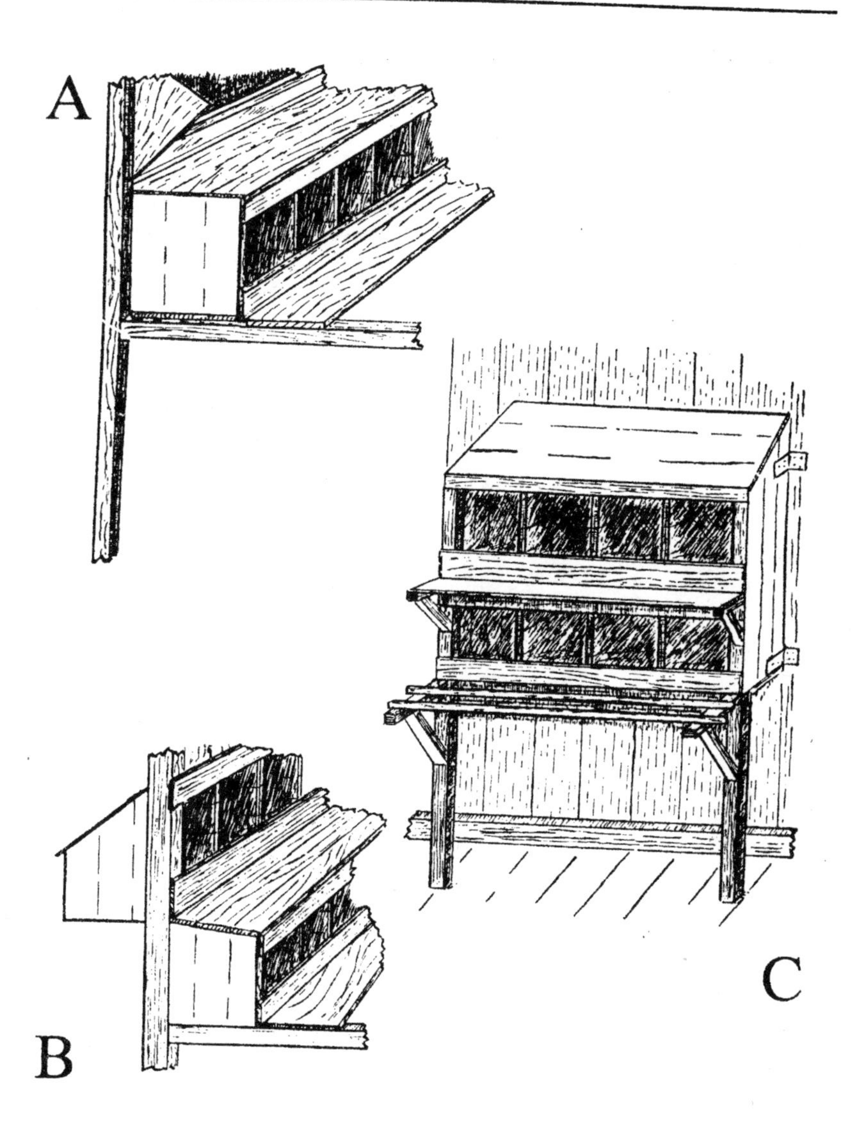

Types of Nestboxes

A = Above floor with stepping tray at front.

B = Inside and outside with access.

C = An 8-compartment unit which can be moved around and for cleaning.

With all nest boxes ensure that birds do not roost on them (if necessary put wire).

6. Give appropriate medication to deal with internal and external parasites. Examples of conditions likely to need attention are as indicated below (more serious health problems may also arise) :

(a) Coccidiosis, (b) Worms of various types ,

(c)Lice , (d) Mite , Fleas and Ticks.

All these can be treated by the poutry manager by the use of a *coccidiostat* or drug for worms or suitable insect powders for the parasites.

Readers are referred to *Poultry Diseases under Modern Management* , Coutts, for details of diagnosis and remedies. They multiply very quickly, if not checked, so early attention is essential.

7. All poor producers should be culled on a regular basis. Hens which lay very few eggs, birds that will not fatten , or chicks that do not grow should all be eliminated and incinerated so that disease does not spread.

8. When very large sheds are involved it may be necessary to bring in contractors to remove accumulated droppings and dirt using high powered jets and suitable disinfectants.

The stock must be looked after on a constant and regular basis; not simply fed and then forgotten, but really cared for just as if they are of the utmost importance–– which they are if part of a commercial business. Any attention which can increase productivity even by a small percentage is worthwhile.

About Soil & Poultry.

A SELECTION OF BREEDS FOR FANCIERS ACCORDING TO THE SOIL ON THEIR FARMS

It is around the amino acid composition that the rations are formulated. Obviously, certain foods contain amino acids and, where necessary to get the correct level of energy, others are added. This is the scientific approach, but it is certainly not the natural approach and cannot be regarded as beneficial to birds or consumers. The purely profit and loss approach without regard to the social consequences must lead to results which are unacceptable to modern society. Keeping birds in cages results in brittle bones which break easily, shortened lives and the possibility of disease on an acute scale. We are not aware of whether poultry have feelings as we know them, but we do know they have distinct **behaviour patterns** and for life to be regarded as normal these should be allowed to be part of their lives*.

An arbitrary decision on the length of life of the hens also seems contrary to human feelings and behaviour. Just because a bird peaks in the first year and then slows down each year thereafter should not be regarded as a good reason to sell them for meat. The useful, profitable life can be up to 3 or 4 years, but more usually is at the end of the second laying season. The author has kept free-range fowl for these periods; the birds lay fewer eggs, yes, but eggs are larger and if culling takes place to remove those which are poor producers, those that remain can still pay for themselves (see Chapter 10 on this aspect).

Natural poultry-keeping should **not** mean, as a matter of course, unprofitable business and in a desire to 'do good' the producer must not shirk from the responsibility of making sure that the flocks of birds are profitable. Culling, trap nesting, record keeping and, if birds are to be kept on a commercial basis, the production of **management accounts**, are all essential requirements.

Besides the various chemical that are added into the food for **intensive production**, there are also colouring additives. The orangy-yellow yolk produced naturally from grass and other natural food on free-range, must be created when the birds are kept in cages without ever seeing the light of day. Without the added colouring the yolks appear a pale yellow, lifeless in form, whereas the free-range egg should glow and appear to be full of health and vitality. The signs of mans' exploitation are all too evident when the facts are examined.

*** See *Behaviour of the Domestic Fowl*, Prof. D G M Wood-Gush, which is a scientific study of behaviour patterns. Available from the publisher.**

THE END RESULT & NATURAL FEEDING

Through scientific advances in food nutrition hens can be made to lay over 300 eggs per year and broiler chicks can be made ready for eating at, say, 8 weeks of age. The question asked is whether the feeding of pellets or mash which have been formulated on the basis of maximum conversion factors is necessarily the best for achieving a **top quality** product.

A recent test on **free range chickens** showed that even they can fall short of the ideal based on what people like to eat. Thus the criteria used were:

1. **Flavour – whether succulent and not too dry.**
2 **Colour of skin and flesh. In this country we prefer white skinned birds.**
3. **Tender without "stringiness"**

Some were found to fall short on excellence. The meat has to be of the correct texture – eatable and digestible with the correct level of moisture (not too wet).

When **intensive** birds are being tasted there is often no flavour and the worst are little better than a form of reconstituted cardboard. Eggs are produced with pale yellow yolks because they have no access to grass.

THE "ORGANIC FOWL"

Much has been said on the definition and merits of the 'organic fowl', but much of this is nonsense. To be a truly organic bird means that rearing and upbringing , as well as the continued management should be along well defined rules. Going back to basics it means that the food used should be grown on organic lines and should be grown without the addition of fertilizers. Different types of grain would be used and presumably there would no objection to natural additives such as cod liver oil , or even mixed into a form of mash after the ingredients have been ground (for easier assimilation).

This is true *Natural* **Poultry Keeping** and an appropriate *standard* is available from the **UK Register of Organic Food Standards.** It seems that although the concept is admirable around 25% extra in prices charged is needed to cover the additional costs. There is resistance in the market to paying this extra amount so the need to educate the consumer or publicize the merits is very apparent.

9

WELFARE OF THE HEN

MAKING LIFE MORE TOLERABLE

In recent years more attention has been paid to the alleged cruelty aspect of keeping birds in cages. In summary form:

1. **Practical Standpoint**

Hens are not humans and do not suffer in cages provided there is adequate room for feeding and exercise. Moreover, the battery system provides eggs at prices the consumer is willing to pay.

2. **Welfare Supporters**

No animal or bird should be restricted too much; all should be free to enjoy green fields, fresh air and unlimited exercise.

Unfortunately, many of the welfare supporters would object very strongly if asked to pay the premium prices expected for free range eggs. Accordingly , in the long run many experts are suggesting some form of compromise will be essential and regulations will ensure that where battery cages are used they comply with specified minimum standards. This is not a sound argument when viewed from a welfare angle, but may be so on economic grounds.

BATTERY HEALTH REQUIREMENTS

Cages for poultry have been the subject of much heated debate and regulations now exist to cover the main requirements. Many of these are common sense measures which look at the welfare of the birds as the main consideration. Inevitably, though, costs will increase and these will have to be covered in the prices charged for eggs.

Many attempts are being made to justify the continuation of the battery system , even to the extent of stating that this is probably healthier than free range. Such allegations are generally without foundation and apply to birds exposed to outdoor conditions when bred for intensive conditions.

An Orchard–type surrounding ; ideal for Growers or Layers

Much natural food is obtainable in these surroundings,but regular 'rest' and liming advisable.

The size of the holding is also of vital importance. In an effort to lay on full services --automatic feeding and water; regulated lighting and other automation -- units of many thousands of birds are installed . This is against the spirit of free range because birds in large units stay indoors near the food conveyors (waiting for them to be turned on) and the fresh air and grass runs are rarely seen. Around 250 pullets per house appears to be the maximum number which should be kept to qualify for the description – 'free range'. However, it must be appreciated that although there is a saving on cages and large, insulated buildings, the costs of providing food and water can be extremely high. Scattering corn for birds improves their health be– cause they have to forage and behave in the normal way as any bird that lives naturally. However, deprived of food and allowed to roam too freely can result in depleted production and possibly lost eggs from birds laying away in hedge bottoms or other secret places.

Scattering corn for a flock of birds of, say, 100, where each is given 3 oz of wheat (a part feed) is a formidable task in itself and this will be considerable where a few thousand birds are kept. A compro– mise may be necessary as discussed in the next section.

RATIONS FOR FREE RANGE

As noted earlier, there are pellets which are formulated for free range and which attempt to make allowance for the natural food such as grass, insects, grit and other items to be found in a field, paddock or orchard.

Alternatively, as stated, mixed corn (wheat, maize and barley) may be used on its own or with pellets provided on an *ad lib* basis. If this combined system is used, the corn may be scattered where the birds have to work to find it. The labour cost is high and for a high level of production, on corn alone, protein content will be too low.

This may be offset by using an automatic feeder developed by breeders of pheasants, where a method was necessary to feed birds periodically without leaving grain available for all and sundry to eat it. The birds will quickly learn to use the facility.

An alternative is to have all–weather out–door hoppers which may be opened at certain times to allow access. Indoor hoppers can also be used, but birds must be *discouraged* from staying indoors all the time.

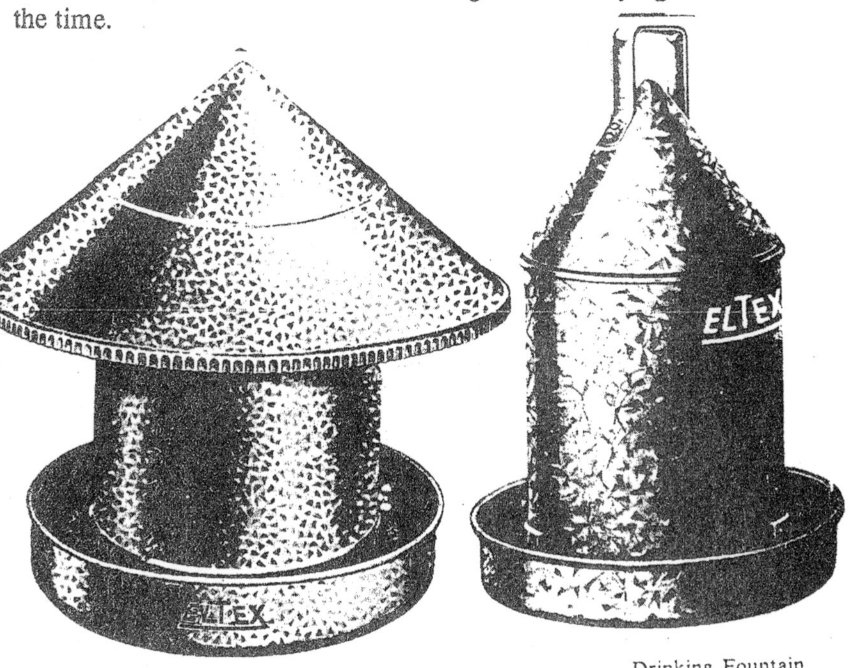

Drinking Fountain

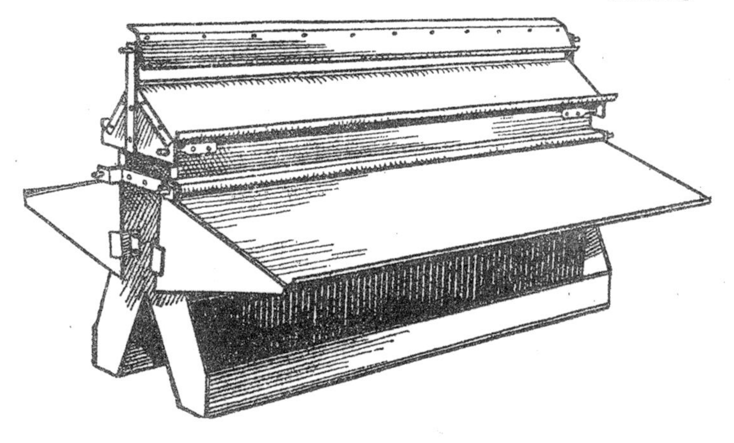

Out–door Hoppers for Free–range.

10

WINTER EGG PRODUCTION

ESSENTIAL REQUIIREMENTS

In order to achieve winter egg production the following facts are important:

1. **A good Winter Laying Breed must be kept***

It is generally conceded that the Mediterranean breeds (Leghorns, Anconas, Andalusians, and Minorcas) are the most prolific egg-layers, but when the calculation is based upon the value of the eggs at the time they are laid, instead of upon the number laid, it is found that there are other breeds more valuable to the egg farmer. This is solely because such breeds can be induced to lay a greater proportion during the late autumn and winter months when eggs are at their top price, whilst the Mediterraneans are somewhat indifferent winter-layers, their season of prolificacy undoubtedly being during spring and summer when eggs are at their cheapest.

Thus Orpingtons, Wyandottes, Rhode Island Reds, Rocks and Croad Langshans - subject to the provisions in the succeeding paragarphs - may all be cited as winter layers, though the claims of Leghorns, Anconas, Andalusians, Campines and Minorcas must not be overlooked if a continuous supply is required.

2. **The Hens must be of a highly productive strain, and bred, if possible, from several generations of winter layers**

Strain counts rather than breed. You may possess Buff Orpingtons that prove themselves excellent winter layers; your next door neighbour, though having the same breed, may find them very indifferent. Strain is defined as: "A race of fowls, which, having been carefully bred by one breeder, or his successors for years, has acquired and established an individual character of its own".

***Reference is to free range production ; the battery system tends to reduce the cyclical effects of variable production in winter. With this section the author has drawn heavily on the experiences of Ralph R. Allen a poultry farmer and author on free range when this was the principal system and standard breeds were kept not hybrids.**

In acquiring new stock, therefore, it is of the utmost importance to purchase only from breeders who definitely make winter egg–production their study.

3. The Pullets which are to be kept for winter layers must be hatched neither too early nor too late

A pullet does not lay until she reaches maturity; the heavier the breed the longer the fowl takes to mature; experience has shown that March hatched birds of the heavy breeds and April hatched birds of the lighter breeds make the best winter layers.

Such are the rules which you must digest, they are, for practical purposes, inflexible.

It is no easy matter to hatch out a large flock of pullets during March, consequently many easy–going poultry keepers are satisfied with their April – or later – efforts. They are foredoomed to disappointment. Because a March hatched Orpington pullet or an April hatched bird of the lighter varieties will probably commence laying in October, pullets of these respective breeds hatched a month later will **not** commence to lay a month later, for their first egg cannot be reckoned upon until about Christmas, or even later.

On the other hand, enthusiasts will often hatch earlier than the time recommended. Certainly their pullets lay earlier, but the number of eggs is usually very limited, after which moulting ensues, and no more eggs are seen till February or March.

There is a correct time to hatch – neither too early nor too late. If you have neglected this necessary precaution, you must purchase the requisite stock from a reputable breeder who has hatched at the correct time.

4. The Hens must not be more than two years old

Poultry will invariably show a profit if expenses are kept low and losses through waste avoided. Again, and again, I see birds on poultry farms that never can be a source of profit to the owner. Deformed birds, sickly birds, old birds, why are they kept? They can never repay their feed–bill, yet alone the cost of labour and housing, whilst they pollute the land equally – sickly birds more so – as the paying hen.

Practical tests have decided that the first laying season of a pullet is the most prolific and profitable; that her next season, though inferior to the first, is still a source of profit, but the returns for her third season, except in individual cases, rarely repay her feed–bill. Get rid of the two–year hen, her carcase is more valuable then than at three, loss is thus avoided and increased profit from her carcase made.

5. The Houses, Yards and other Appliances must be so laid out and constructed as to ensure comfort

Draught–proof and water–proof houses are imperative, but the former does not imply lack of ventilation. Personally I strongly recommend open–fronted houses, even during winter months, which can be either partially or wholly closed according to climatic conditions. They should, if possible, have a sunny aspect. Inattention to these details has for years been greatly responsible for the paucity in the winter egg supply, and although the majority pay greater attention to the housing accommodation, improvement is still to be desired on many poultry plants. A final word on this subject:

Do not overcrowd – overcrowding, particularly in winter, is responsible for much disease; the birds become overheated during the night and their stamina is weakened when exposed to wintry blasts in the morning.

6. The Food must contain a sufficiently large proportion of those elements which are necessary for the Formation of Eggs, the Repair of Tissue, and the Production of Meat

It is no mere coincidence that during late spring and summer there is an abundant supply of insect life (animal food) together with prolonged hours of daylight for the hens to search for it, and that at these seasons they are more prolific in egg–production than at others.

I unhesitatingly assert that it is cause and effect. Consequently, if you would have eggs out of season you must supply that which nature withholds; ie, extra protein and greens.

Other chapters in this book deal with general feeding, so that at the moment it is unnecessary to go into this subject in detail, but a few additional remarks as to how to attain the best results from feeding will be opportune.

SEASONAL DISTRIBUTION OF EGG PRODUCTION

A USA Experiment

The data on which the laying results are based are the trap nest records of Barred Plymouth Rocks some years ago involving detailed monthly egg records of more than 2,400 birds, collected in a period of nine consecutive years.

The chief results of this analysis were:

1. The mean or average monthly egg production exhibits the following characteristic changes in the course of the laying year :

(a) The lowest mean production of the year is the month of November.

(b) There were average monthly production increases in December and January at a relatively very rapid rate.

(c) There was a slackening up in the rate of increase in February, which probably represents the point of the ending of the first, or winter, cycle of egg-production. This February slackening up amounts in many cases to an actual decrease in productiveness as compared with the point attained in January.

(d) The mean production reaches a maximum in March.

(e) While the mean production for April is practically the same as that for March, there is a steady decline after April on to the end of the laying year in October.

(f) There is a tendency toward a slightly larger drop in mean production in May. This is the period of natural broodiness.

2. The present data indicate that only a trifle more than a quarter of the total eggs produced are laid in the winter part of the year (November 1st to February 28th).

Average production for the year was 128 eggs which is poor by today's standards.

British Comparison

The results shown are based on the actual records of an experiment conducted at a Welsh poultry farm. The following salient facts emerged:

(a) That October appears to be our month for the lowest mean average production of the year.

(b) That in November and December the mean monthly production increases.

(c) That there is a slacking in January and February, which probably represents the point of the ending of the first, or winter, cycle of egg-production, and which, in the case of Buff Orpingtons, may be intensified by their aggravated inclination to broodiness.

(d) The mean production reaches a maximum in March.

(e) After March there is a decline until October, the end of the laying year, the decline being twice broken to a small extent, which would probably be caused by climatic conditions or a varying degree of broodiness.

(f) Again there is agreement with American deductions, the tendency to a larger drop in May - the period of natural broodiness - being apparent.

(g) Whilst in America, 28% of the total output was laid during the winter third of the year (November 1st to February 28th), only 23% was obtained from the Welsh birds.

(h) That the average for the year attained by the Amercian birds was 128 eggs, whilst the Welsh Buffs averaged $152^1/_3$.

Surveying the question broadly, therefore, there is but little variation in the seasonal distribution of eggs between America and Britain the slight divergences probably being accounted for by climatic differences.

What is apparent is that anyone keeping birds on free range must do all possible to combat the ill effects of the low months. Drops in production due to the moult must be expected, but it will be necessary to increase the protein content of the food to help to grow the new feathers. When the birds go to roost, the lighting may be introduced to give 12 hours light (with natural daylight). Some may argue that this is not Natural Poiultry-keeping, but we all use and benefit from light supplied, and, with poultry, it is essential to give sufficient light to allow the hens to eat sufficient to produce eggs.

A method of combatting a shortage of eggs is to have pullets available and coming into lay in October so that they start to peak when the hens are in one of the low periods. Agreed, eggs will be smaller at first, but will improve.

LAYING TRIALS

When the eggs were produced by many poultry farmers there was a high level of competition to produce the best layers from a specific breed. With the high degree of mechanization into factory farming the hen was no longer a bird which needed to be put into a competition; what would be the point when the hybrids produced on computer information came from a very few producers anyway. The farmer, in his belief that he would earn a larger profit abandoned the standard breeds and went for the hybrids, which were usually kept in cages.

What had not been appreciated was that the suppliers of birds were to be limited in numbers and, eventually, would be reduced to one or two only. The reduction in body size and the endless pursuit to produce more eggs would inevitably increase the danger of disease. If birds were to be bred from the same source over a very long period then any weakness would show itself. Moreover, the high level of stocking would increase the dangers of any infection spreading.

At one point in the history of poultry keeping the rearing of pullets was regarded as being more acceptable if in reasonable batches and outdoors on fresh ground to avoid disease. With the intensive system even the chicks and pullets would be reared intensively so the birds would never see proper daylight; agreed, some did use other methods, but gradually these were reduced in numbers.

Feathering, bone strength, digestion, exercise, and many other factors do not function properly under artificial conditions so, although the controls in mammoth houses were improved, they could not simulate natural conditions and eventually the birds had to suffer. Only when the farmer meets disaster, such as losing thousands of birds when fans break down, or there is a break out of disease, is any action taken to relieve the suffering which occurred. Even when the poor birds reach the end of their very short lives (usually 2/3 years of age) the torture is still carried on; they are bundled into crates and sent to a different kind of factory where they end up as food for humans or animals.

The fact that poultry are kept primarily for food must not be overlooked and therefore in the end they must be killed, but this should be done in a dignified manner, with the minimum of discomfort to the birds.

The laying competitions that took place with standard bred stock showed that very good results could be obtained, provided the birds selected had constitutional and physical vigour. The breeds entered for these closely controlled competitions were varied, and included were Golden Wyandottes (200 eggs), White Wyandottes (276 eggs), Buff Rocks (216), Rhode Island Red (285 eggs). Moreover, the trap nesting system used showed which birds were not laying and, since the pullets were kept in separate pens, any mortality could be recorded.

It is readily acknowledged that the nutriment derived by the hen from its food is utilised primarily for the repair of waste tissue and maintenance of heat, and only after these vital processes are provided for is the surplus – if any – devoted to egg–production. Hence it becomes a question of nutriment, which in other words means digestion, for when the digestive organs are working in perfection, then, and then only, can the hen derive the maximum amount of nutriment from her rations. The content of the food is of vital importance ; amino acids, vitamins and other essentials in terms of protein and calcium are vital.

The laying competition food consisted of natural foods such as follows:

1. Grain feed in the morning, scattered in the litter. Also fed in the evenings.

Wheat, cracked maize, clipped oats in the proportions 2 : 2 : 1.

2. Mash ad lib feeding

Middlings, 35 to 45 per cent
Bran, 15 to 20 per cent.
Sussex Ground Oats, 15 to 20 per cent.
Maize Meal, 15 to 20 per cent.
Fish Meal, 7 to 10 per cent.

When the grass in the runs became in short supply Alfalfa meal was added to the mash.

This feeding was nothing special so the results came from the ability of the stock to lay and the management. Since the methods used came very close to what is regarded as Natural Poultry–keeping it should be apparent that the system can work.

11
ESTABLISHING WHETHER POULTRY FARMING IS FEASIBLE

FACTORS TO CONSIDER

Essential questions to ask are:

1. What experience have you had of this particular business you are about to enter upon?

2. Are you possessed of sound constitution and business intelligence, and demonstrated an aptitude to be successful in your career?

3. Have you the necessary capital to commence business?

1. What experience have you had of this particular business you are about to enter upon?

This question cannot be satisfactorily answered by the great majority of those who engage in poultry–farming, and here is the first and most prolific cause of failure.

An apprenticeship must be served in poultry–farming; you must, to be successful, understand the business from A to Z, in exactly the same way as in other vocation in life. This may be acquired, firstly and preferably, by actual residence and work on a farm for an extended period, and not for a course of a few weeks; or, secondly, by retaining your present calling and keeping a few birds as a hobby, gradually increasing your stock with your knowledge, until you are possessed of a farm capable of affording a living.

2. Are you possessed of sound constitution and ordinary business intelligence, and have you shown some degree of success in business?

* This Appendix is for those who contemplate a career in poultry farming.

Poultry–farming is ceaseless routine from early morning till dark, and at certain seasons even later; clerical work, posting of books and answering correspondence to be performed during the evening, and this after not only a long but also a hard day's work on the farm – that is why sound health is essential.

What about the business intelligence? The combination of experience and hard work will raise chickens, but what are you going to do for a market? If you are not possessed of ordinary business intelligence you may not be able to sell the produce at a price to repay what you have spent on them.

3. Have you the necessary Capital to commence the Business?

Repeatedly I have seen the combination of all required by questions one and two come to disaster a few months after starting, solely because the business was inadequately capitalised and there is difficulty in meeting the food and labour costs.

With poultry–farming immediate profits are not made: there are the eggs to produce, to hatch, the chicks to grow, to fatten and finally to market, before there is a return, and though I believe poultry–farming to be a legitimate business which can be made profitable, I do not mean that it is an El Dorado which will return 100 per cent on capital invested.

I am not going to raise here the vexed question of the actual amount, or the minimum amount, necessary to commence, so much depends upon the aspirant's experience, his tastes and style of living, his aptitude for work and creative genius with tools; but this I say deliberately, that in my opinion, there should at least be ample to pay:

1. For the cost of the initial scheme decided upon:

(a) Size of flock;

(b) Sheds and other accommodation;

(c) Feeding stuffs;

(d) Costs for labour, electricity, transport and other essentials.

A Cash Flow statement, showing when money will be expected to come in and go out will be vital. Banks may help and so will the suppliers of food, but it is safer to rely on your recources for banks often want the loan back at the time when it is most needed and show little sympathy when difficulties arise.

2. For the maintenance of the owner (and those dependant upon him) for twelve months; and

3.. A reserve of certainly £10,000 or more* , to meet those unexpected expenses that were never thought of. Without this, anybody embarking is running a risk that cannot be warranted.

If the three questions can be satisfactorily answered, then my belief is that a living can be made at the business, but if one essential is lacking, then the risk becomes most hazardous.

If money cannot be found then it is better to keep a few poultry and enjoy the hobby, rather than risk all on a venture that might fail because of lack of capital.

New Hampshire Reds

*Note: This is a rough guide at the present time; inflation and related factors will vary this figure. Adequate capital must be considered in relation to the overall size of the operation. Preparation of a Business Plan and a Cash Flow Budget will be essential requirements, especially if the intention is to be fairly substantial and obtain an overdraft or loan.

12

CONCLUSIONS

Recognizing the Need

The book has covered the many aspects of keeping poultry at different levels. The discerning reader will appreciate that, in effect, implementing Natural Poultry Keeping is like turning back the clock 50 years or more to a period before the Intensive systems were imported from the USA. The effect of these systems resulted in fewer people being needed to run a poultry farm, but more capital was required to house birds under intensive conditions. Battery cages, automatic feed and water equipment, fully insulated houses with an environment controlled by fans and heaters, and many other requirements were all brought into production. There was a farming revolution and little attention was paid to the welfare needs of the laying stock unless it affected output.

There followed the introduction of 'hybrid' types of poultry, birds which ate a relatively small amount, but produced as many eggs as possible. The birds would be kept for one year – two at the most – and then new stock had to be penned.

Foodstuff research was directed at the needs of the hybrid breeds so that the maximum results could be obtained. The scientific approach replaced the skilled management needed formerly. The food supplied the needs of the specific birds, whether layers. breeders, broilers or chicks.

It was the "egg factory" which replaced the normal way of keeping birds. Birds became units of production which created on the basis of what they were fed; chemicals which increased egg production were introduced. Whether these chemicals will affect humans in some adverse way is not clear, but it does seem quite clear that they could **not** be regarded as natural requirements.

The Battery System
Now should be phased out

Free Range – where the birds live naturally

The Alternatives

Natural Poultry Keeping should eliminate any doubts because the birds are eating what is a natural part of their diet. Moreover, they are not being kept in close confinement, like in a prison cell, never seeing the light of day, never scratching for food, or being fed on a chemical based diet; all factors which cannot be looked upon as desirable or humane.

The products from the intensive methods cannot be regarded with favour in terms of what is best for the consumer. Additives to make the egg yolk deeper in colour and to strengthen the shell cannot be regarded as acceptable. The law on additives to food generally is quite strict and artificial colouring must be controlled. Why should the poor hen be given dyes and other additives to make the egg, one of our staple foods, appear more attractive.

All the natural goodness should be brought back into the egg by the acceptable method of feeding natural foods in natural surroundings; in fields, paddocks, orchards and back gardens, where the birds can see the normal daylight, feel the sunshine and live a life that is acceptable in all ways.

RECAP

We have to recognize that the turning back of the clock will not be easy. At the same time, there should be acceptance that the move is both essential and inevitable. Cruelty to birds cannot be tolerated in any way so the intensive systems, which have been shown to fall short of the basic requirements, must be discarded. The costs are said to be higher under free range, but this is open to question. Admittedly. in the Winter months the production will suffer, but this can be offset by careful selection of new stock, introduced at the correct time, and by restricting the time outside when the weather is really inclement.

The movement towards free range in its entirety or to semi–free range (fully in the Summer and the Barn system in Winter) must go on for the benefit of the birds as well as the consumer, who could well be affected by any virus which lurks in eggs from diseased birds.

Birds must have access to fresh air and grass, and areas where scratching can take place. The foodstuff should consist of various types of corn, insects, grass, and other greenstuff, grit (soluble and insoluble for digestion), and an ample supply of fresh water; all a natural diet.

The domesticated fowl has been part of mans' environment for thousands of years and has provided food during that time. We must treat it in an humane fashion, thus allowing its useful work to continue.

INDEX (NPK)